LaTeX Graphics with TikZ

A practitioner's guide to drawing 2D and 3D images, diagrams, charts, and plots

Stefan Kottwitz

BIRMINGHAM—MUMBAI

LaTeX Graphics with TikZ

Copyright © 2023 Packt Publishing

All rights reserved. No part of this book may be reproduced, stored in a retrieval system, or transmitted in any form or by any means, without the prior written permission of the publisher, except in the case of brief quotations embedded in critical articles or reviews.

Every effort has been made in the preparation of this book to ensure the accuracy of the information presented. However, the information contained in this book is sold without warranty, either express or implied. Neither the author, nor Packt Publishing or its dealers and distributors, will be held liable for any damages caused or alleged to have been caused directly or indirectly by this book.

Packt Publishing has endeavored to provide trademark information about all of the companies and products mentioned in this book by the appropriate use of capitals. However, Packt Publishing cannot guarantee the accuracy of this information.

Group Product Manager: Alok Dhuri

Publishing Product Manager: Akshay Dani

Senior Editor: Kinnari Chohan

Technical Editor: Jubit Pincy

Copy Editor: Safis Editing

Project Coordinator: Manisha Singh

Proofreader: Safis Editing

Indexer: Subalakshmi Govindhan

Production Designer: Joshua Misquitta

Developer Relations Marketing Executive: Deepak Kumar and Mayank Singh

Business Development Executive: Puneet Kaur

First published: June 2023

Production reference: 1260523

Published by Packt Publishing Ltd.

Livery Place

35 Livery Street

Birmingham

B3 2PB, UK.

ISBN 978-1-80461-823-3

www.packtpub.com

To Till Tantau, the inventor of TikZ, and to Christian Feuersänger, the inventor of pgfplots. With thanks to Henri Menke, the current maintainer of TikZ, and Mark Wibrow for his contributions.

Further thanks go to Kjell Magne Fauske for creating TeXample.net, Izaak Neutelings for working on TikZ.net, and Denis Bitouzé, Patrick Bideault, and Alain Matthes for supporting TikZ.fr.

– Stefan Kottwitz

Contributors

About the author

Stefan Kottwitz studied mathematics in Jena and Hamburg. He works as a network and IT security engineer for Lufthansa Industry Solutions.

He has been offering LaTeX support on internet forums for many years. He maintains the web forums LaTeX.org and goLaTeX.de and the question and answer (Q&A) sites TeXwelt.de and TeXnique.fr. He runs the TeX graphics gallery sites TeXample.net, TikZ.net, and PGFplots.net, the TeXlive.net online compiler, the TeXdoc.org service, and the CTAN.net software mirror.

A moderator of the TeX Stack Exchange site and matheplanet.com, he publishes ideas and news from the TeX world on his blogs LaTeX.net and TeX.co.

He has also authored the *LaTeX Beginner's Guide* in 2011, the second edition in 2021, and the *LaTeX Cookbook* in 2015, all by Packt.

About the reviewers

Izaak Neutelings got his master's and PhD degrees at the University of Zurich (UZH). Now he works at the CMS experiment at CERN, where he does fundamental research in the field of experimental particle physics, hunting for new particles in proton collisions. He has written lecture notes for introductory physics courses at UZH, fully illustrated with TikZ figures.

Joseph Wright is the author of several widely used LaTeX packages and is a member of the LaTeX Project team and the author. Joseph is a chemist by training, and in his day job is a university lecturer in inorganic chemistry.

Table of Contents

Preface — xiii

1
Getting Started with TikZ — 1

Technical requirements	1	With a vanilla TeX distribution	5
What is TikZ?	2	With an operating system TeX installation	5
Alternative graphics packages	2	Installing from sources	6
The LaTeX picture environment	2	**Working with the TikZ documentation**	**6**
MetaPost	3		
Asymptote	3	**Creating our first TikZ figure**	**7**
PSTricks	4	**Summary**	**8**
Benefits of TikZ	4	**Further reading**	**8**
Installing TikZ	5		

2
Creating the First TikZ Images — 9

Technical requirements	9	Using relative coordinates	18
Using the tikzpicture environment	10	Using units	19
Working with coordinates	12	Drawing geometric shapes	20
Cartesian coordinates	13	Using colors	21
Polar coordinates	15	Summary	23
Three-dimensional coordinates	16	Further reading	23

3
Drawing and Positioning Nodes — 25

Technical requirements	25
Understanding nodes	26
Using shapes and anchors	28
A rectangle shape	29
The circle and ellipse shapes	30
The coordinate shape	30
More shapes	31
Spacing within and around nodes	34
Positioning and aligning nodes	36
Using anchors and relative positioning	36
Placing nodes along a line	38
Aligning nodes at the text baseline	39
Aligning whole pictures at a node text baseline	40
Adding labels and pins	43
Putting images into nodes	44
Summary	46
Further reading	47

4
Drawing Edges and Arrows — 49

Technical requirements	49
Connecting nodes by edges	50
Adding text to edges	51
Diving deeper into edge options	54
Path options	55
Connection options	55
Drawing arrows	56
Mathematical arrow tips	57
Barbed arrow tips	57
Geometric arrow tips	58
Customizing arrow tips	58
Using the to operation	59
Summary	61
Further reading	61

5
Using Styles and Pics — 63

Technical requirements	63
Understanding styles	63
Defining and using styles	64
Inheriting styles	68
Using styles globally and locally	69
Giving arguments to styles	71
Creating and using pics	73
Summary	76
Further reading	76

6

Drawing Trees and Graphs — 79

Technical requirements	79	Positioning in a matrix	95
Drawing trees	80	Summary	98
Creating mind maps	88	Further reading	99
Producing graphs	91		

7

Filling, Clipping, and Shading — 101

Technical requirements	101	Reverse clipping	114
Filling an area	102	Shading an area	117
Understanding the path interior	102	Axis shading	117
		Radial shading	120
The nonzero rule	103	Ball shading	121
The even odd rule	107	Bilinear interpolation	121
Comparing the nonzero rule and the even odd rule	108	Color wheel	122
		Summary	125
Clipping a drawing	110	Further reading	125

8

Decorating Paths — 127

Technical requirements	127	Replacing paths with ticks	135
Pre- and post-actions for using a path multiple times	128	Decorating paths with text	137
		Adding markings	138
Understanding decorations	129	Adjusting decorations	139
Exploring the available decoration types	132	Summary	142
		Further reading	143
Morphing paths	132		

9

Using Layers, Overlays, and Transparency — 145

Technical requirements	146	Positioning pictures on the background of a page	157
Using transparency	146	Summary	160
Drawing on background and foreground layers	153	Further reading	160
Overlaying LaTeX content with TikZ drawings	155		

10

Calculating with Coordinates and Paths — 161

Technical requirements	161	Counting loop repetitions	170
Repeating in loops	162	Evaluating the loop variable	171
Calculating with coordinates	165	Remembering the loop variable	171
Adding and subtracting coordinates	165	Calculating intersections of paths	172
Computing points between coordinates	166		
Projecting on a line	167	Summary	175
Adding angles	168	Further reading	176
Evaluating loop variables	170		

11

Transforming Coordinates and Canvas — 177

Technical requirements	177	Transforming the canvas	187
Shifting nodes and coordinates	178	Summary	188
Rotating, scaling, and slanting	179	Further reading	188

12

Drawing Smooth Curves — 189

Technical requirements	189	Using a smooth plot to connect points	193
Manually creating a smooth curve through chosen points	190	Specifying cubic Bézier curves	195

Using Bézier splines to connect given points	196	Summary	201
Using the Hobby algorithm for smoothly connecting points	197	Further reading	202

13

Plotting in 2D and 3D — 203

Technical requirements	204	Calculating plot intersections	215
Introducing plotting	204	Adding a legend	216
Creating and customizing Cartesian axes, ticks, and labels	206	Using the polar coordinate system	217
Understanding axis environments	207	Parametric plotting	218
Customizing ticks and labels	209	Plotting in three dimensions	220
Using plotting commands and options	211	Summary	223
Filling the area between plots	213	Further reading	223

14

Drawing Diagrams — 225

Technical requirements	225	Line charts	240
Creating flowcharts	226	Bar charts	243
Linear flow diagrams	226	Pie charts	247
Circular flow diagrams	231	Wheel charts	249
Building relationship diagrams	232	Summary	250
Writing descriptive diagrams	237	Further reading	251
Producing quantitative diagrams	239		

15

Having Fun with TikZ — 253

Technical requirements	253	Meeting the TikZlings	256
Drawing cute creatures	254	Building snowmen	258
Playing with rubber ducks	254	Playing with penguins	259

Playing and crafting	263	Building with bricks	266
Creating jigsaw puzzles	264	**Drawing world flags**	270

Index 273

Other Books You May Enjoy 282

Preface

LaTeX Graphics with TikZ is a practical introduction to producing graphics in LaTeX. It features TikZ, a powerful modern computer graphics package. This book will help you write mathematical, scientific, or technical papers with graphics. The book guides you through the initial challenges and provides a rapid learning process. Even though using an external graphics editor may seem like a more accessible option at first sight, it will turn out that learning TikZ is more than worth the effort.

This book starts with essential topics such as installing TikZ and learning the fundamental syntax. It offers step-by-step examples that begin with understanding coordinate systems, drawing geometric shapes, and working with nodes, anchors, edges, and arrows. You will also learn to utilize styles to produce consistent graphics easily while saving typing work.

Furthermore, this book covers clipping, filling, shading, and adding decorations. You will learn about calculations with coordinates and transformations of coordinates and canvas.

This book will help you create professional-looking diagrams and plots in two and three dimensions for visualizing your ideas and data.

With *LaTeX Graphics with TikZ* to hand, you can quickly start with TikZ and enjoy its many benefits.

Who this book is for

If you're a LaTeX user in school, academia, or industry, and you are looking to add figures such as diagrams, plots, and graphics in general to your thesis, articles, or any document, this book offers a practical and fast-paced introduction to producing such figures. Whether you're a student, teacher, or engineer, this book is highly beneficial. Once you have experience in LaTeX or have read any LaTeX beginner's book or tutorial, you can successfully work with this book.

What this book covers

Chapter 1, Getting Started with TikZ, introduces TikZ. It discusses alternative graphics packages and emphasizes TikZ's benefits. You'll thoroughly understand what TikZ is all about and its unique philosophy. You'll receive guidance on installing TikZ, and you will walk through creating a small drawing. Additionally, you will get helpful tips for accessing TikZ's and other packages' documentation.

Chapter 2, *Creating the First TikZ Images*, walks you through creating a LaTeX document with a drawing from scratch. You will gain a solid understanding of the TikZ syntax and learn about cartesian and polar coordinates in two and three dimensions. Additionally, you'll learn how to create basic geometric shapes and incorporate color into your designs.

Chapter 3, *Drawing and Positioning Nodes*, introduces the fundamental concept of nodes. You'll learn how to draw nodes in various shapes, position and align them, and add text, images, and labels.

Chapter 4, *Drawing Edges and Arrows*, shows how to connect nodes by edges, straight and curvy lines, and arrows. You'll see how to add text labels on the edges and adjust alignment, position, and orientation. You'll learn to use line styles and customized arrow tips in one or both directions.

Chapter 5, *Using Styles and Pics*, teaches you how to define and apply global and local styles for TikZ elements. You will learn how to use styles on nodes and edges and apply them to entire pictures or selected parts of a picture using scopes. Additionally, you will learn about using mini TikZ pictures as building blocks.

Chapter 6, *Drawing Trees and Graphs*, guides you through creating tree structures to depict parent-child relationships hierarchically. It shows how to draw mind maps to visualize ideas and introduces a concise syntax for generating graphs. Additionally, this chapter offers a practical technique for arranging objects in a matrix format similar to LaTeX's tabular environment.

Chapter 7, *Filling, Clipping, and Shading*, starts with more advanced techniques. You'll learn how to fill complex paths, clip pictures to specific areas, and add shading that transitions smoothly from one color to another.

Chapter 8, *Decorating Paths*, introduces techniques for adding creative effects to lines and curves, such as making them wavy, zigzag, or bumpy. You'll also learn how to print text along a curved path and apply multiple actions on a single path.

Chapter 9, *Using Layers, Overlays, and Transparency*, demonstrates how to create drawings on different layers, allowing you to place objects behind text or images. You will learn how to use transparency to improve this effect. Additionally, you will discover how to superimpose TikZ annotations on top of regular LaTeX text and add background images to document pages, similar to watermarks.

Chapter 10, *Calculating with Coordinates and Paths*, shows the efficient way of letting TikZ calculate coordinate values. This chapter covers coordinate calculation, distance and projection calculation, and calculating intersections of paths. You'll also discover how to save time and streamline your code by using loops to repeat commands.

Chapter 11, *Transforming Coordinates and Canvas*, focuses on shifting, rotating, and scaling nodes and coordinates using transformations. You'll learn skills that enable you to make precise adjustments and repositioning, whether you need to make minor tweaks or complex changes to your drawings.

Chapter 12, *Drawing Smooth Curves*, explores different methods to draw easy curves smoothly with gentle slopes, smooth transitions, and without sharp corners or spikes, similar to freehand-like drawings.

Chapter 13, *Plotting in 2D and 3D*, deals with visualizing data in a coordinate system. It covers customizing Cartesian and polar axes and adding legends, plotting explicit and parametric functions in 2D and 3D, calculating plot intersections, and filling between plots.

Chapter 14, *Drawing Diagrams*, shows how to create flowcharts, relationship diagrams, descriptive diagrams, and quantitative diagrams. The emphasis is on using packages to generate whole diagrams in a more automated way.

Chapter 15, *Having Fun with TikZ*, showcases examples of how skilled TikZ users enjoyed programming add-on packages and sharing them with the TikZ community. You'll see how to draw cute animals, human shapes, nation flags, and game pieces.

To get the most out of this book

For using TikZ, a TeX installation, such as TeX Live, MiKTeX, or MacTeX, is required on your computer. TikZ and LaTeX are compatible with most operating systems, including Windows, Linux, macOS, and other Unix operating systems. All code examples in this book have been tested with TeX Live 2023 on Debian Linux and with MacTeX 2023 on macOS Ventura. For those who do not wish to install LaTeX, code examples are available on `https://tikz.org`, which includes an online compiler that makes the code accessible also for smartphone and tablet users. Alternatively, you can register on `https://overleaf.com` to compile the examples obtained from GitHub or TikZ.org.

TikZ version 3.1.9.a has been used to develop and test the code examples in this book. All references to sections in the manual refer to that version. A future version may have a different section numbering.

If you are using the digital version of this book, we advise you to type the code yourself or access the code from the book's GitHub repository (a link is available in the next section). Doing so will help you avoid any potential errors related to the copying and pasting of code.

Download the example code files

Throughout the book, concise code snippets are used to explain concepts without repetitive LaTeX document body and preambles. The entire code is available online for reference and further exploration.

All examples use the `standalone` class. You can use the example codes and TikZ in general in any LaTeX document class.

You can download the example code files for this book from GitHub at `https://github.com/PacktPublishing/LaTeX-graphics-with-TikZ`. If there's an update to the code, it will be updated in the GitHub repository.

You can open the entire code bundle as a single project on Overleaf using the following link: `https://www.overleaf.com/docs?snip_uri=https://tikz.org/code.zip`.

We also have other code bundles from our rich catalog of books and videos available at `https://github.com/PacktPublishing/`. Check them out!

Download the color images

We also provide a PDF file that has color images of the screenshots and diagrams used in this book. You can download it here: `https://packt.link/7hkX1`

Conventions used

There are a number of text conventions used throughout this book.

`Code in text`: Indicates code words in text, database table names, folder names, filenames, file extensions, pathnames, dummy URLs, user input, and Twitter handles. Here is an example: "Write `\draw [blue] circle (1cm);` to get a blue circle."

A block of code is set as follows:

```
\begin{tikzpicture}
   \draw (-0.5,0) to ["text"] (0.5,0);
\end{tikzpicture}
```

When we wish to draw your attention to a particular part of a code block, the relevant lines or items are set in bold:

```
\begin{axis}[axis lines=center]
   \addplot {x^3/5 - x};
\end{axis}
```

Any command-line input or output is written as follows:

```
$ texdoc tikz
```

Bold: Indicates a new term, an important word, or words that you see onscreen. For instance, words in menus or dialog boxes appear in **bold**. Here is an example: "On the right-hand side, we see the so-called **Transpose** of the matrix."

> **Tips or important notes**
> Appear like this.

Get in touch

Feedback from our readers is always welcome.

TikZ and LaTeX questions: If you have any questions about TikZ, LaTeX, or this book, you can post them at the author's forum at `https://latex.org`

General feedback: If you have questions about any aspect of this book, email us at `customercare@packtpub.com` and mention the book title in the subject of your message.

Errata: Although we have taken every care to ensure the accuracy of our content, mistakes do happen. If you have found a mistake in this book, we would be grateful if you would report this to us. Please visit `www.packtpub.com/support/errata` and fill in the form.

Piracy: If you come across any illegal copies of our works in any form on the internet, we would be grateful if you would provide us with the location address or website name. Please contact us at copyright@packt.com with a link to the material.

If you are interested in becoming an author: If there is a topic that you have expertise in and you are interested in either writing or contributing to a book, please visit `authors.packtpub.com`.

Share your thoughts

Once you've read *LATEX Graphics with TikZ*, we'd love to hear your thoughts! Scan the QR code below to go straight to the Amazon review page for this book and share your feedback.

`https://packt.link/r/1804618233`

Your review is important to us and the tech community and will help us make sure we're delivering excellent quality content.

Download a free PDF copy of this book

Thanks for purchasing this book!

Do you like to read on the go but are unable to carry your print books everywhere?

Is your eBook purchase not compatible with the device of your choice?

Don't worry, now with every Packt book you get a DRM-free PDF version of that book at no cost.

Read anywhere, any place, on any device. Search, copy, and paste code from your favorite technical books directly into your application.

The perks don't stop there, you can get exclusive access to discounts, newsletters, and great free content in your inbox daily

Follow these simple steps to get the benefits:

1. Scan the QR code or visit the link below

https://packt.link/free-ebook/9781804618233

2. Submit your proof of purchase
3. That's it! We'll send your free PDF and other benefits to your email directly

1
Getting Started with TikZ

First, congratulations on using LaTeX! You have already chosen an excellent tool for writing, and now you are ready to add high-quality figures to your documents.

To work with this book, you should have a good understanding of LaTeX and know how to work with a LaTeX editor and compiler. If you are still learning about LaTeX, here are two recommended books:

- The *LaTeX Beginner's Guide, Second Edition* by Packt Publishing gives you a fast start; you can find more information about it at `https://latexguide.org`
- The *LaTeX Cookbook* by Packt Publishing offers many ready-to-use examples for various document types. Look at the book's website at `https://latex-cookbook.net`

This chapter shall be a quickstart. We will cover the following topics:

- What is TikZ?
- Alternative graphics packages
- Benefits of TikZ
- Installing TikZ
- Working with TikZ documentation
- Creating our first TikZ figure

We will briefly look at the technical requirements in each chapter, so let's do that now.

Technical requirements

You need to have a LaTeX distribution installed on your computer, for example, TeX Live (`https://tug.org/texlive`), MacTeX (`https://tug.org/mactex`), or MiKTeX (`https://miktex.org`). A full installation is recommended. It may take up to 8 GB of space on your hard disk, but you don't need to worry about missing packages.

Alternatively, you can use Overleaf (`https://overleaf.com`). It's an excellent online LaTeX editor and compiler; that's why it requires a permanent internet connection to be able to use it.

All code examples of this chapter are available on GitHub at `https://github.com/PacktPublishing/LaTeX-graphics-with-TikZ/tree/main/01-Getting-Started-with-TikZ`.

At `https://tikz.org`, you can also find all code examples from this book. You can edit and compile to PDF directly on that website, so you can even use a smartphone or tablet to work with this book's code.

What is TikZ?

The inventor of TikZ, Till Tantau, created the name as a recursive acronym in German. **TikZ** stands for **TikZ ist kein Zeichenprogramm**, which translates to *TikZ is not a drawing program*. It's Tantau's jokey way of emphasizing that you cannot expect to draw with it like with a pen or just mouse clicks, such as with Microsoft Paint on Windows, Paintbrush on a Mac, Adobe Illustrator Draw, or the free Inkscape vector graphics editor.

Simply said, TikZ is a set of TeX commands for drawing graphics. Just like LaTeX is code that describes a document, TikZ is code that describes graphics and looks like LaTeX code. With TikZ, you write `\draw [blue] circle (1cm);` to get a blue circle with a 1 cm radius in your PDF document.

The origin of TikZ is called **PGF**, which stands for **Portable Graphics Format** and is a set of graphics macros that can be used with pdfLaTeX and the classic DVI/PostScript-based LaTeX. Today, we consider TikZ as the frontend and PGF as the backend. So, to install TikZ, we need to look for *pgf* in the TeX package manager. From time to time, we will see a command with `pgf` in the name, but as authors and not developers, we will write in TikZ syntax almost exclusively.

Alternative graphics packages

Before we start, let's quickly look at where we come from and what else is out there.

The LaTeX picture environment

LaTeX itself defines some basic graphics commands. We can use a `picture` environment for this.

To get an idea of how it works, let's have a quick look at a minimal example:

```
\setlength{\unitlength}{1cm}
\begin{picture}(1,1)
  \put(0,0){\circle{1}}
  \put(-0.5,0){\line(1,0){1}}
```

```
    \put(-0.3,0.06){text}
\end{picture}
```

The output is the following:

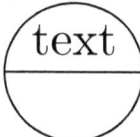

Figure 1.1 – A classic LaTeX picture drawing

Here, we did the following:

1. We set a base length. All numbers in drawing commands are seen as multiples of this base unit length.
2. We used `\put(x,y){...}` to put something at the Cartesian coordinate position x, y.
3. We wrote `\circle{x}` to get a circle with a diameter of x times the unit length.
4. We used `\line(x,y){z}` to get a line in vector direction (x,y) with a length of z times the unit length.

There are a few more commands, such as for drawing arrows and ovals, but that's pretty much it. Now comes the fun part: lines, circles, and ovals are taken from unique fonts, so a drawing is put together from symbols. Consequently, lines can just have some predefined slope values, and circles are available with just some diameter values up to about 14 mm. The drawing is approximated and doesn't look perfect. That was the time LaTeX was invented. Today, there's the `pict2e` package, which extends the classic picture environment a bit and mainly removes some restrictions but plays in the same league. If you are further interested in the basic picture mode, look at https://texdoc.org/pkg/pict2e.

MetaPost

MetaPost is a drawing language of its own, powerful and mature. It produces PostScript code that can be converted to PDF. MetaPost is an external program or library. It was an early graphics companion for TeX users and is still in use. Its syntax differs from LaTeX so we won't cover it further in the book. Visit https://metapost.eu for more information.

Asymptote

Asymptote is a vector graphics language inspired by MetaPost. It is very mathematically oriented, with actual 3D capabilities. It can use LaTeX for typesetting labels in its drawings, so the images are consistent with the LaTeX document. However, note that it is external software, which is why we won't be discussing it anymore in this LaTeX book, but you may visit https://asymp.net for more details.

PSTricks

PSTricks is an extensive TeX macro collection for producing PostScript code that can be converted to PDF. LaTeX editors can do that conversion automatically with the help of support packages. The PSTricks package is large in size and extremely powerful, and many additional packages make use of it. If we did not have TikZ, PSTricks would be *the* way to go with LaTeX. Still, there are the restrictions of having to use PostScript, difficult syntax, and less user base support compared to TikZ. So, over time, TikZ became more popular. Visit `https://pstricks.org` for more information.

Now that we have had a quick look at other graphics packages, let's see what TikZ offers compared to the alternative packages in this section and classic GUI software.

Benefits of TikZ

Compared to classic drawing programs where you click with the mouse on shapes and toolbars and drag and drop graphic elements, TikZ is very different. With TikZ, you *program graphics with code*.

That means your graphics will be the following:

- **Precise**: You get the exact placement of graphic and text elements using anchors, baselines, alignment, relative positioning, and implicit coordinate calculations.
- **Consistent**: TikZ blends in perfectly with LaTeX. You can use LaTeX fonts, symbols, formulas, colors, and macros within your drawing, and your drawing details will precisely match your LaTeX document design. That would be different if you imported some externally made images.
- **High-quality**: TikZ generates scalable PDF images that look fine when you zoom in or out. There are no blurry or pixelated images.
- **Efficient**: Similar figures mean similar code and similar styles; it's all reusable, and when you adjust global styles, you change the appearance of all corresponding figures in your document. This means there's less need to repeat things.
- **Cross-platform**: Your drawing will work with every major operating system and online compiler that runs LaTeX. Even more, you can use all common (La)TeX engines, such as pdfLaTeX, XeLaTeX, LuaLaTeX, and even classic plain TeX and ConTeXt, a big macro package and interesting alternative to LaTeX.

With TikZ, you inherit the benefits of LaTeX regarding scientific typesetting, quality, separation of styles and content, and version control, such as GitHub.

Many developers have created packages on top of TikZ for the easier creation of diagrams and charts, plots, trees, and other types of images with a more accessible interface. And there is a large user base that has put thousands of TikZ drawings with complete code on the internet into browsable galleries, such as `https://tikz.net` and `https://texample.net`.

It's good practice for beginners to browse a TikZ gallery, choose an example that roughly matches their desired result, and use that code as a starting point. By reading this book, you will be able to understand such code and modify it. The excellent – but 1,300 page-long – TikZ manual can then be your reference for looking up specific styles.

Now that we've teed up, let's get into the details of installing TikZ.

Installing TikZ

As you already have LaTeX installed, you just need to ensure that the `pgf` and `xcolor` packages are installed. You can install them in three main ways, which are discussed in the following sub-sections.

With a vanilla TeX distribution

If you installed LaTeX from DVD or via the internet from an original TeX distribution, use its package manager to install the `pgf` and `xcolor` packages. These are the three principal TeX distributions, along with installation details:

- **TeX Live**: Start TeX Live Manager (`tlmgr` or `tlshell`), then search and install the `pgf` and `xcolor` packages
- **MiKTeX**: Use the MiKTeX package manager (`mpm`) to install `pgf` and `xcolor`
- **MacTeX**: Use TeX Live Utility to install `pgf` and `xcolor`

When you want to update TikZ later on, run your regular TeX distribution updates, and `pgf` will be updated as well.

With an operating system TeX installation

If your LaTeX installation comes from your operating system repositories, which is usually the case with Linux, you should use your operating system tools. For example, to install via the command line in a terminal session, perform the following, depending on your operating system:

- **Debian**: Depending on your OS version, run the following:

    ```
    aptitude install pgf
    ```

    Or, the following:

    ```
    apt-get update
    ```

    And then run this:

    ```
    apt-get install latex-xcolor
    apt-get install pgf
    ```

Or, run this:

```
apt-get install texlive-pictures
```

The latter contains some more graphics-related LaTeX packages.

- **Ubuntu**: This is like Debian, but you should use the prefix `sudo` before commands.
- **Redhat, CentOS, Fedora**: For these, you can use `yum`:

```
sudo yum makecache
sudo yum -y install texlive-pgf
```

Or, use `dnf`:

```
sudo dnf makecache
sudo dnf -y install texlive-pgf
```

TikZ (PGF) will also be updated when you update your Linux version.

Installing from sources

This is rare and usually not needed, but experts may feel adventurous, so let's quickly mention two ways:

- You can download a **TeX Directory Structure** (**TDS**)-compliant TikZ zip file (`pgf.tds.zip`) from `https://ctan.org/pkg/pgf`. The TikZ manual describes the installation in the *Installation in a texmf Tree* section.
- You can visit the TikZ GitHub project site at `https://github.com/pgf-tikz/pgf` to download and install it as described there.

But seriously, let's stick with the TikZ package with either your TeX distribution or your operating system for consistency and compatibility.

Working with the TikZ documentation

The LaTeX and TikZ installations contain documentation. You can access it in the following two ways:

- **Windows**: Run the `cmd` app via the **Start** menu
- **Apple macOS or Linux**: Open a terminal window

Then, type this command:

```
texdoc tikz
```

Your PDF viewing app will open and display the TikZ manual in all its 1,300-page size and glory. But don't feel intimidated, for the following reasons:

- It's good to have so many features so well documented in a huge reference with an extensive index, hyperlinks, full-text search, and of course, a lot of examples.
- Hundreds of pages are about the PGF backend, the basic layer, and the system layer on the driver level. You don't need that as a user.
- It describes all additional libraries and utilities.
- It contains five tutorials.

I hope that in the future, this manual will be split into a TikZ manual, a PGF backend reference for developers, and tutorials.

If you don't have `texdoc` or the documentation on your computer, such as when you use the Overleaf online compiler, you can open the manual at `https://texdoc.org/pkg/tikz` and download it to your computer.

An exciting alternative is at `https://tikz.dev`: that's the TikZ manual in HTML format produced using the `lwarp` package. Especially on smartphones, such a reflowing document is much more readable than a PDF document with a fixed paper size.

With all the setup done and all the important points discussed, let us move on and create our first TikZ figure.

Creating our first TikZ figure

Our first goal is to create a TikZ drawing that is the same as *Figure 1.1*, which we made in the classic LaTeX picture mode, to get a feeling of the TikZ basics.

To be able to use TikZ, you need to perform the following three steps:

1. Load the `tikz` package in your document preamble:

    ```
    \usepackage{tikz}
    ```

2. TikZ provides additional features with separate libraries. Here, we load the `quotes` library for adding annotations with an easy quoting syntax that we will use in the drawing:

    ```
    \usetikzlibrary{quotes}
    ```

3. Use a `tikzpicture` environment for the drawing. The first code snippet we saw in this chapter, for the picture environment, will look like this with TikZ:

    ```
    \begin{tikzpicture}
      \draw circle (0.5);
    ```

```
    \draw (-0.5,0) to ["text"] (0.5,0);
\end{tikzpicture}
```

This results in the following output:

Figure 1.2 – Our first TikZ drawing

We draw a circle with a radius of 0.5 cm at the default origin, that is (0,0). Then we draw a line from (-0.5,0) to (0.5,0) in Cartesian coordinates, with the label `text`.

Well done, that's your first drawing! Now you know the basic steps for drawing with TikZ.

Summary

In this chapter, we learned what TikZ is, how to install it, and how to access the documentation. We had a first glimpse at the syntax and created our first TikZ figure of the book. Rest assured, there are many more to come!

In the coming chapters, we will learn more about TikZ commands, objects, and styles, to create more fancy drawings. In the next chapter, we'll learn about the essential toolbox, with coordinates, paths, colors, lines, circles, and other curves and shapes.

Further reading

The TikZ manual is an excellent and comprehensive reference book. We will refer to it many times. You can find the manual at `https://texdoc.org/pkg/tikz` in PDF format and at `https://tikz.dev` in handy reflowable HTML sections, which makes it easier to read on a smartphone or tablet.

The *LaTeX Graphics Companion* by Michel Goossens, Frank Mittelbach, et al. is a comprehensive book about creating LaTeX graphics. With the first edition published in 1997 and the second and latest edition published in 2007, it doesn't cover the newest developments, even though there's a reprint of the second edition from 2022. When I bought the reprint and noticed that TikZ is completely missing in that book, I decided to write the book you hold in your hand now.

PSTricks: Graphics and PostScript for TeX and LaTeX by Herbert Voss is a great reference book specifically about PSTricks with many examples.

MetaPost: Grafik für TeX und LaTeX by Walter Entenmann is a very recommendable book about Metapost. Unfortunately, it is only available in the German language.

2
Creating the First TikZ Images

This chapter will work with the most basic but essential concepts.

Specifically, our topics are as follows:

- Using the `tikzpicture` environment
- Working with coordinates
- Drawing geometric shapes
- Using colors

This gives us the foundation to move on to more complex drawings in the upcoming chapters.

It's good if you already know the basics of geometry and coordinates, but we will have a quick look at the parts we need.

By the end of this chapter, you'll learn how to create colored drawings with lines, rectangles, circles, ellipses, and arcs and how to position them in a coordinate system.

Technical requirements

You need to have LaTeX on your computer, or you can use Overleaf to compile the code examples of this chapter online. Alternatively, you can go with the book's website, where you can open, edit, and compile all examples. You can find the code for this chapter at `https://tikz.org/chapter-02`.

The code is also available on GitHub at `https://github.com/PacktPublishing/LaTeX-graphics-with-TikZ/tree/main/02-First-steps-creating-TikZ-images`.

Using the tikzpicture environment

In the previous chapter, we saw that we basically load TikZ and then use a `tikzpicture` environment that contains our drawing commands. Let's go step by step to create a document that will be the base of all our drawings in this chapter. Our goal is to draw a rectangular **grid** with dotted lines. Such a grid is really beneficial in positioning objects in our pictures later on. I usually start with such a helper grid, make my drawing, and take the grid out in the final version of the drawing.

As it's one of our first TikZ examples, we will do it step by step and then discuss how it works:

1. Open your LaTeX editor. Start with the `standalone` document class. In the class options, use the `tikz` option and define a border of 10 pt:

   ```
   \documentclass[tikz,border=10pt]{standalone}
   ```

2. Begin the document environment:

   ```
   \begin{document}
   ```

3. Next, begin a `tikzpicture` environment:

   ```
   \begin{tikzpicture}
   ```

4. Draw a thin, dotted grid from the coordinate (-3,-3) to the coordinate (3,3):

   ```
   \draw[thin,dotted] (-3,-3) grid (3,3);
   ```

5. To better see where the horizontal and vertical axis is, let's draw them with an arrow tip:

   ```
   \draw[->] (-3,0) -- (3,0);
   \draw[->] (0,-3) -- (0,3);
   ```

6. End the `tikzpicture` environment:

   ```
   \end{tikzpicture}
   ```

7. End the document:

   ```
   \end{document}
   ```

Compile the document and look at the output:

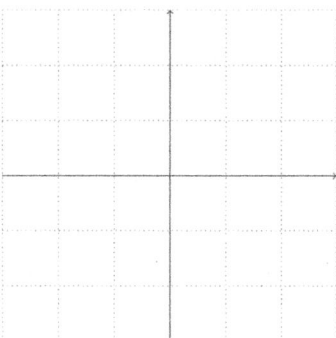

Figure 2.1 – A rectangular grid

In *step 1*, we used the `standalone` document class. That class allows us to create documents that consist only of a single drawing and cuts the PDF document to the actual content. Therefore, we don't have an A4 or letter page with just a tiny drawing, plus a lot of white space and margins.

To get a small margin of 10 pt around the picture, we wrote `border=10pt` because, with a small margin, it looks nicer in a PDF viewer. Since the `standalone` class is designed for drawings, it provides a `tikz` option. As we set that option, the class loads TikZ automatically, so we don't have to add `\usepackage{tikz}` anymore.

After we started the document in *step 1*, we opened a `tikzpicture` environment in *step 3*. Every drawing command will happen in this environment until we end it. As it's a LaTeX environment, it can be used with optional arguments. For example, we could write `\begin{tikzpicture}[color=red]` to get everything we draw in red unless we specify otherwise. We will talk about valuable options later in this book.

Step 4 was our main task of drawing a grid. We used the `\draw` command that we will see exceptionally often throughout this book. We specified the following:

- **How**: We added `thin` and `dotted` options in square brackets because that's the LaTeX syntax for optional arguments. So, everything the `\draw` command does will now be in thin and dotted lines.
- **Where**: We set (-3,-3) as the start coordinate and (3,3) as the end coordinate. We will look thoroughly at the coordinates in the next section.
- **What**: The `grid` element is like a rectangle where one corner is the start coordinate, to the left of it, and the other corner is the end coordinate, to the right of it. It fills this rectangle with a grid of lines. They are, as we required before, thin and dotted.

`\draw` produces a **path** with coordinates and picture elements in between until we end with a semicolon. We can sketch it like the following:

```
\draw[<style>] <coordinate> <picture element> <coordinate> ...
;
```

Every path must end with a semicolon. Paths with coordinates, elements, and options can be pretty complex and flexible – the rule to end paths with a semicolon allows TikZ to parse and understand where such paths end and other commands follow.

The lines in a grid have a distance of 1 by default. The optional `step` argument can change that. For example, you could write `grid[step=0.5]` or do that right at the beginning as the `\draw` option, such as the following:

```
\draw[thin,dotted,step=0.5] <coordinate>
  <picture element> <coordinate> ... ;
```

In *step 5*, we have drawn two lines. The picture element here is a straight line between the coordinates given. We use the convenient `--` shortcut that stands for a line. The `->` style determines that we shall have an arrow tip at the end. In the next section, we will draw many lines.

Finally, we just ended the `tikzpicture` and `document` environments.

> **TikZ, document classes, and figures**
>
> In this book, we will focus on TikZ picture creation. Remember that we can use TikZ with any LaTeX class, such as `article`, `book`, or `report`. Furthermore, TikZ pictures can be used in a `figure` environment with `label` and `caption`, just like `\includegraphics`.

While this section showed a manageable number of commands, we should have a closer look at the concept of coordinates, which is now the topic of our next section.

Working with coordinates

When we want TikZ to place a line, a circle, or any other element on the drawing, we need to tell it where to put it. For this, we use **coordinates**.

Now, you may remember elementary geometry from school or have looked at a good geometry book. In our case, we will use our knowledge of geometry mainly to position elements in our drawings.

Let's start with classic geometry and how to use it with TikZ.

Cartesian coordinates

You may remember the Cartesian coordinate system you learned in school. Let's quickly recap it. In the two dimensions of our drawing, we consider an *x* axis in the horizontal direction going from left to right and a *y* axis in the vertical order going from bottom to top. Then, we define a point by its distance to each axis. Let's look at it in a diagram:

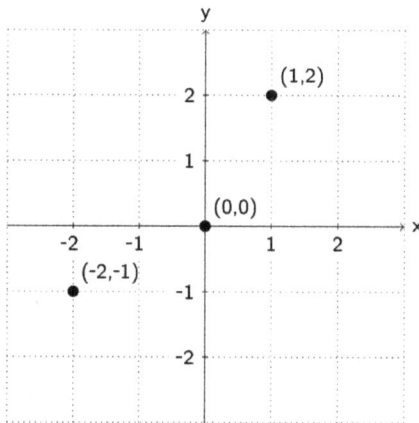

Figure 2.2 – Cartesian coordinate system

In *Figure 2.2*, we see a point **(0,0)** that we call the **origin**. It has a distance of zero to each axis. Then there's the point, **(1,2)**, that has a distance to the origin in a positive *x* direction of **1** and a positive *y* direction of **2**. Similarly, for the **(-2,1)** point, we have an *x* value of **-2**, since it goes in the negative direction, and a *y* value of **-1** for the same reason.

Labels at the *x* axis and *y* axis and a grid help us to see the dimensions. We will reuse the grid from *Figure 2.1* when we next draw lines.

Remember, we draw elements between coordinates, and -- is the code for a line. So, the following command draws a line between the (2,-2) and (2,2) coordinates:

```
\draw (2,-2) -- (2,2);
```

We can add more coordinates and lines to this command – let's make it a square. And to better see it over the grid, let's make it have very thick blue lines:

```
\draw[very thick, blue] (-2,-2) -- (-2,2)
    -- (2,2) -- (2,-2) -- cycle;
```

Here, `cycle` closes the path, so the last line returns to the first coordinate.

The full context – that is, the complete LaTeX document with the `cycle` command – is highlighted in the code for *Figure 2.1*:

```
\documentclass[tikz,border=10pt]{standalone}
\begin{document}
\begin{tikzpicture}
  \draw[thin,dotted] (-3,-3) grid (3,3);
  \draw[->] (-3,0) -- (3,0);
  \draw[->] (0,-3) -- (0,3);
  \draw[very thick, blue] (-2,-2) -- (-2,2)
    -- (2,2) -- (2,-2) -- cycle;
\end{tikzpicture}
\end{document}
```

When you compile this document, you get this picture:

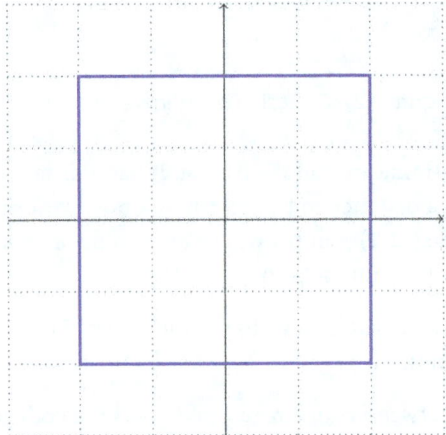

Figure 2.3 – A square in Cartesian coordinates

We used the `\draw` command to put lines at and between coordinates. How about something else? In TikZ, we can draw a circle with a certain radius as an element, with that radius as an argument in parentheses, such as `circle (1)` with a radius of 1. Let's replace the `--` lines with that and remove the now unnecessary `cycle`, and the command now looks like this:

```
\draw[very thick, blue] (-2,-2) circle (1) (-2,2)
    circle (1) (2,2) circle (1) (2,-2) circle (1);
```

Compile, and you get this in the PDF document:

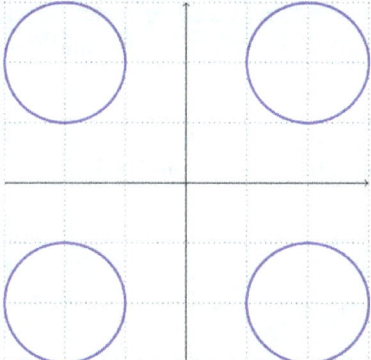

Figure 2.4 – Circles in Cartesian coordinates

This example emphasizes how we use the \draw command – as a sequence of coordinates with picture elements at those coordinates. As you saw, we can draw several elements in a single \draw command.

With Cartesian coordinates, it was easy to draw a square. But how about a pentagon? Or a hexagon? Calculating corner coordinates looks challenging. Here, angle- and distance-based coordinates can be more suitable; let's look at this next.

Polar coordinates

Let's consider the same plane as we had in the last section. Just now, we define a point by its distance to the origin and the angle to the *x* axis. Again, it's easier to see it in a diagram:

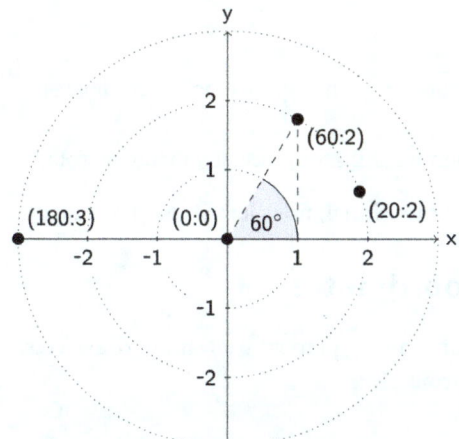

Figure 2.5 – Polar coordinate system

We have a point with the polar coordinates **(60:2)**, which means a distance of 2 from the origin with an angle of **60** degrees to the *x* axis. TikZ uses a colon to distinguish it from Cartesian coordinates in polar coordinate syntax. The syntax is (angle:distance). So, **(20:2)** also has a distance of **2** to the origin, **(0:0)**, and an angle of 20 degrees to the *x* axis, and **(180:3)** has a distance of 3 and an angle of 180 degrees.

Now, it becomes easier to define points for a hexagon – we specify the angles in multiples of 60 degrees, and all have the same distance from the origin, **(0:0)**; let's choose 2. Our drawing command becomes as follows:

```
\draw[very thick, blue] (0:2) -- (60:2) -- (120:2)
    -- (180:2) --(240:2) -- (300:2) -- cycle;
```

With the same grid code in the LaTeX document from the previous sections, we get this result from compiling:

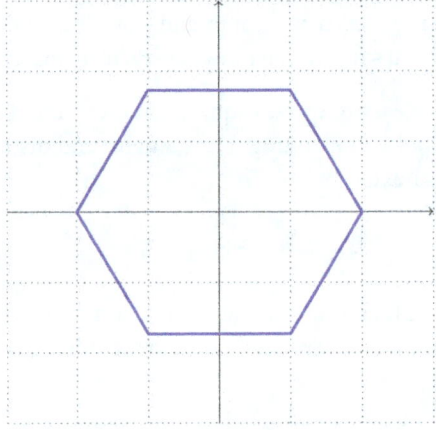

Figure 2.6 – A hexagon in polar coordinates

Polar coordinates are handy when we think of points by distance, rotation, or direction.

Until now, everything was two-dimensional; now, let's step up by one dimension.

Three-dimensional coordinates

We could use a projection on our drawing plane if we want to draw a cube, a square, or spatial plots. The most famous is **isometric projection**.

TikZ provides three-dimensional coordinate systems and options. Here is a quick view of how we can use them:

- Specify *x*, *y*, and *z* coordinates that shall be the projection of our three-axis vectors:

    ```
    \begin{tikzpicture}[x={(0.86cm,0.5cm)},
      y={(-0.86cm,0.5cm)}, z={(0cm,1cm)}]
    ```

- Use three coordinates now. We will draw the same square as in *Figure 2.3*, with 0 as the *z* value, so still in the *xy* plane:

    ```
    \draw[very thick, blue] (-2,-2,0) -- (-2,2,0)
      -- (2,2,0) -- (2,-2,0) -- cycle;
    ```

For a better view, we shall again draw axes, as shown in *Figure 2.3*. Furthermore, we add a circle with a radius of 2. With the necessary aforementioned code highlighted, the full code example is as follows:

```
\documentclass[tikz,border=10pt]{standalone}
\begin{document}
\sffamily
\begin{tikzpicture}[x={(0.86cm,0.5cm)},
  y={(-0.86cm,0.5cm)}, z={(0cm,1cm)}]
  \draw[very thick, blue] (-2,-2,0) -- (-2,2,0)
    -- (2,2,0) -- (2,-2,0) -- cycle;
  \draw[->] (0,0,0) -- (2.5, 0,   0) node [right] {x};
  \draw[->] (0,0,0) -- (0,   2.5, 0) node [left]  {y};
  \draw[->,dashed] (0,0,0) -- (0,   0, 2.5) node [above] {z};
  \draw circle (2);
\end{tikzpicture}
\end{document}
```

This gives us a skewed view, where the axes and circle help in recognizing it as a 3D isometric view:

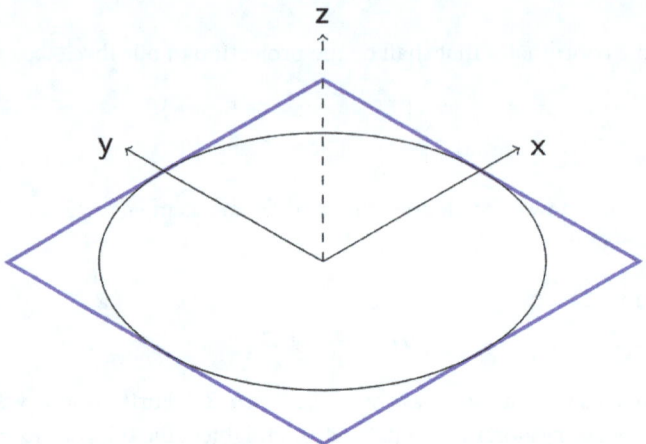

Figure 2.7 – The square and circle in three dimensions

In later chapters, we will work with additional libraries and packages for three-dimensional drawing.

Until now, we have used only absolute coordinates, which refer to the origin and axes. How about a reference to another point, with a distance or angle? We will now look at that.

Using relative coordinates

When we use \draw with a sequence of coordinates, we can state the relative position to the first coordinate by adding a + sign. So, +(4,2) means the new coordinate is plus 4 in the *x* direction and plus 2 in the *y* direction. Note that with +, it is always relative to the first coordinate in this path section.

Let's try this in our code with the grid from *Figure 2.3*:

```
\draw[very thick, blue] (-3,-1) -- +(1,0)
  -- +(2,2) -- +(4,2) -- +(5,0) -- +(6,0);
```

Compile, and you get the following:

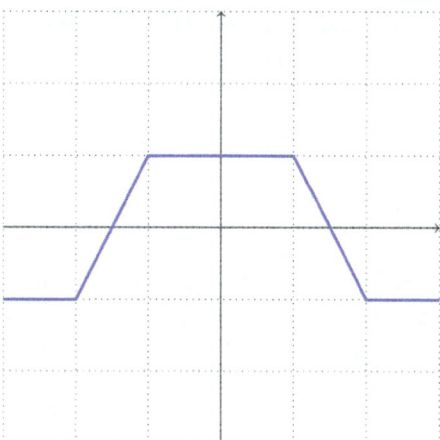

Figure 2.8 – Drawing with relative coordinates

That's not so handy – always looking back to the first coordinate. Luckily, TikZ offers another syntax with double plus signs. For example, ++(1,2) means plus one in the x direction and plus 2 in the y direction, but from the previous point. That means we can move step by step.

The modified drawing command for the same output is as follows:

```
\draw[very thick, blue] (-3,-1) -- ++(1,0)
    -- ++(1,2) -- ++(2,0) -- ++(1,-2) -- ++(1,0);
```

We get the same drawing as shown in *Figure 2.8*; currently, it's much easier to follow the movement from one coordinate to the next. That's why this syntax is pretty popular. Remember that -- here is not the negative version of ++; it's the line element. The use of -- ++ together can look confusing, but they are two different things – a line and a relative positioning modifier.

Using units

You may already have wondered what a coordinate, (1,2), or a radius of 2 can mean in a document regarding the size of the PDF. Mathematically, in a coordinate system, it's clear, but in a document, we need actual width, height, and lengths.

So, by default, 1 means 1 cm. You can use any LaTeX dimension, so you can also write (8mm,20pt) as a coordinate or (60:1in) for 60 degrees with a 1-inch distance.

You can change the default unit lengths of 1 cm to any thing else you like. If you write `\begin{tikzpicture}` `[x=3cm,y=2cm]` you get $x = 1$ as 3 cm, and $y = 1$ will be 2 cm. So, `(2,2)` would mean the point, `(6cm, 4cm)`. It's an easy way of changing the dimensions of a complete TikZ drawing. For example, change *x* and *y* to be twice as big in the `tikzpicture` options to double a picture in size.

We have now seen how to draw lines, circles, and a grid. Let's look at more shapes now.

Drawing geometric shapes

We want to progress from high-speed to advanced TikZ concepts, so let's have a compact summary of what we can draw in this basic setting – that is, we start with `\draw <coordinate>` (that's the current coordinate) and continue with some of the following elements:

- **Line**: `-- (x,y)` draws a line from the current coordinate to (x,y).
- **Rectangle**: `rectangle (x,y)` draws a rectangle where one corner is the current coordinate, and the opposite corner is (x,y).
- **Grid**: Like `rectangle` but with lines in between as a grid.
- **Circle**: `circle (r)` was a short syntax we used previously, but the extended syntax is `circle [radius=r]`, which draws a circle with the center at the current coordinate and a radius of r.
- **Ellipse**: `ellipse [x radius = rx, y radius = ry]` draws an ellipse with a horizontal radius of rx and a vertical radius of ry. The short form is `ellipse (rx and ry)`.
- **Arc**: `arc [start angle=a, end angle=b, radius=r]` gives a part of a circle with a radius of r at the current coordinate, starting from angles *a* to angles *b*. The short command version is `arc (a:b:r)`.

  `arc [start angle=a, end angle=b, x radius=rx, y radius=ry]` gives a part of an ellipse with an *x* radius of rx and a *y* radius of ry at the current coordinate, starting from angle *a* and going to angle *b*. The short syntax would be `arc (a:b:rx and ry)`.

Let's have a few examples to see what these commands do:

1. Draw a circle with a radius of 2 at the origin:

   `\draw (0,0) circle [radius=2];`

2. Next, draw an ellipse with a horizontal radius of `0.2` and a vertical radius of `0.4`:

   `\draw (-0.5,0.5,0) ellipse [x radius=0.2, y radius=0.4];`

3. Now, draw the same ellipse at `(0.5,0.5)`:

   `\draw (0.5,0.5) ellipse [x radius=0.2, y radius=0.4];`

4. Next, draw an arc that looks like a smile:

   ```
   \draw (-1,-1) arc [start angle=185, end angle=355,
     x radius=1, y radius=0.5];
   ```

5. Finally, draw a rectangle with the lower-left corner at -3,-3 and the upper-right corner at 3,3:

   ```
   \draw (-3,-3) rectangle (3,3);
   ```

When you use all the commands from *steps 1 to 5* in a `tikzpicture` environment and compile, you get the following:

Figure 2.9 – A smiley in a rectangle

This result of the command examples still looks a bit dull. Let's improve it a bit and fill it with color.

Using colors

We can add colors as options to `\draw`, as we did for *Figure 2.3* when we added blue lines. When we look at circles, ellipses, and rectangles, we can see that the element can have one color while the inner area can have another color. We can add the latter using the `fill` option.

It's easier to see it with an example – to draw a blue circle filled with yellow. For this, we can write the following:

```
\draw[blue,fill=yellow] (0,0) circle [radius=2];
```

Let's now fill colors in *Figure 2.9*. We'll use `fill=yellow` for the circle, `fill=black` for the ellipses, and make the arc thicker by using `very thick`. Also, let's omit the rectangle. Our commands are as follows, in a complete document, with the changes highlighted:

```
\documentclass[tikz,border=10pt]{standalone}
\begin{document}
\begin{tikzpicture}
```

```
    \draw[fill=yellow] (0,0) circle [radius=2];
    \draw[fill=black] (-0.5,0.5,0)
      ellipse [x radius=0.2, y radius=0.4];
    \draw[fill=black] (0.5,0.5,0)
      ellipse [x radius=0.2, y radius=0.4];
    \draw[very thick] (-1,-1) arc [start angle=185,
      end angle=355, x radius=1, y radius=0.5];
  \end{tikzpicture}
\end{document}
```

When we compile this document, we get the following:

Figure 2.10 – A smiley with color

TikZ has another way of filling called **shading**. Instead of filling with a uniform color, shading fills an area with a smooth transition between colors. For our smiley, we chose a predefined `ball` shading that gives a three-dimensional impression. We set the `shading=ball` and `ball color=yellow` options for the face, and `ball color=black` for the eyes. The code becomes the following:

```
\draw[shading=ball, ball color=yellow] (0,0)
  circle [radius=2];
\draw[shading=ball, ball color=black] (-0.5,0.5,0)
  ellipse [x radius=0.2, y radius=0.4];
\draw[shading=ball, ball color=black] (0.5,0.5,0)
  ellipse [x radius=0.2, y radius=0.4];
\draw[very thick] (-1,-1) arc [start angle=185,
  end angle=355, x radius=1, y radius=0.5];
```

Now, our four draw commands produce an even fancier smiley:

Figure 2.11 – A smiley with a three-dimensional appearance

In *Chapter 7, Filling, Clipping, and Shading*, we will learn more about choosing and mixing colors and explore various ways of filling areas with colors.

Summary

In this chapter, we got used to the basic TikZ syntax, and we learned to draw with different kinds of coordinates. We saw how to draw lines, rectangles, grids, circles, ellipses, and arcs, and how to color them.

Combining text and shapes with alignment options is even more important and worthwhile. That's the concept of nodes, which we will explore in the next chapter.

Further reading

The TikZ manual includes some excellent tutorials in *Part I, Tutorials and Guidelines*. You can find the manual at `https://texdoc.org/pkg/tikz` in PDF format and `https://tikz.dev/tutorials-guidelines`.

Coordinates and coordinate systems are explained in depth in *Part III, Section 13, Specifying Coordinates*, and online at `https://tikz.dev/tikz-coordinates`.

The geometric shapes we learned to draw in this chapter are called **path operations** in the TikZ manual. *Part III, Section 14, Syntax for Path Specifications*, is the reference for them. You can read that section online at `https://tikz.dev/tikz-paths`.

3
Drawing and Positioning Nodes

Text elements of TikZ pictures are called **nodes**. This feature gives you excellent control over placing and arranging text in graphics, and you can combine it with additional drawing elements.

In this chapter, you will learn how to draw nodes with various shapes containing text and how to position them.

We will deal with the following topics:

- Understanding nodes
- Using shapes and anchors
- Positioning and aligning nodes
- Adding labels and pins
- Spacing within and around nodes
- Putting images into nodes

By the end of this chapter, you will be ready to draw your first images with text elements.

Technical requirements

You need either a local LaTeX installation on your PC or an online compiler such as Overleaf or the book's website compiler. You can find all code examples for this chapter at `https://tikz.net/contents/chapter-03`.

The code is also available on GitHub at `https://github.com/PacktPublishing/LaTeX-graphics-with-TikZ/tree/main/03-drawing-and-positioning-nodes`.

Again, we sometimes just show code snippets to not spend too much book space on repetitive commands, such as `\begin{tikzpicture}` and `\end{tikzpicture}`. Every code snippet is available as a fully compilable document at `TikZ.org` and GitHub; you can use those code examples for exercises.

In this chapter, we will use the `shapes` and `positioning` libraries included in TikZ and the `tikzpeople` and `enumitem` packages.

Understanding nodes

In TikZ, a node is a piece of text that can have a specific **shape**. By default, nodes have a rectangular shape, but we can choose between many other shapes, such as circles, ellipses, polygons, stars, clouds, and many more. Using shapes other than rectangles and circles requires loading the `shapes` library. So, from now on, we will add this line to our TikZ documents:

```
\usetikzlibrary{shapes}
```

Let's start with elementary examples. We can place a simple piece of text on the coordinates x=4 and y=2 with the following command:

```
\draw (4,2) node {TikZ};
```

It gives us just the word `TikZ` at the position (4,2). When we want TikZ to also draw the border, we add the `draw` option to the node:

```
\draw (4,2) node[draw] {TikZ};
```

We can choose a border color, fill it with a color, and choose a text color, for example:

```
\draw (4,2) node[draw, color=red, fill=yellow, text=blue] {TikZ};
```

What started as simple text now looks like this:

Figure 3.1 – A node with colors

The following are rules of thumb to note:

- The node text is in curly braces and is always required
- Coordinates are in parentheses
- Design options are in square brackets

Since we use nodes very often, there is the `\node` command for drawing them.

Let's take the following command:

```
\draw (4,2) node [draw] {TikZ};
```

We could write the following command instead:

```
\node [draw] at (4,2) {TikZ};
```

We can give nodes names. We use parentheses for this. Let's create three nodes: a rectangle node (`r`), a circle node (`c`), and an ellipse node (`e`):

```
\node (r) at (0,1)    [draw, rectangle]  {rectangle};
\node (c) at (1.5,0)  [draw, circle]     {circle};
\node (e) at (3,1)    [draw, ellipse]    {ellipse};
```

This gives us the following picture:

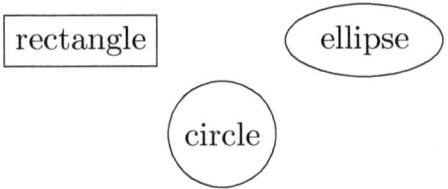

Figure 3.2 – Nodes with different shapes

We can use these names for later drawings. For example, now we can add arrows from one node to another, using compass directions, such as `north`, `south`, `east`, `west`, and others:

```
\draw[->] (r.east)   -- (e.west);
\draw[->] (r.south)  -- (c.north west);
\draw[->] (e.south)  -- (c.north east);
```

We get three arrows, as shown in the following picture:

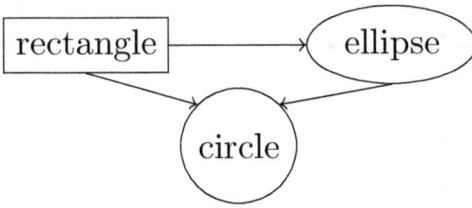

Figure 3.3 – Nodes with arrows

These compass directions are called **anchors**. That's because we can use them to anchor a node on a position. The best way is to see it in a picture. Let's put a red-filled circle at (4,2) and then add a rectangular node:

```
\draw[fill=red] (4,2) circle[radius=0.1];
\node at (4,2) [draw, rectangle] {rectangle};
```

We see the following output:

Figure 3.4 – Default anchor

You can see that the rectangle node is placed in a way that its center is at the given coordinate of (4,2). If we want to have (4,2) as the southwest corner, we can define this corner as the anchor of the node:

```
\node at (4,2) [draw, rectangle, anchor=south west]
    {rectangle};
```

Together with the red-filled circle, it now looks like this:

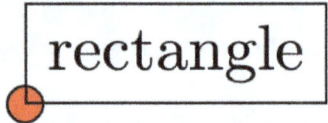

Figure 3.5 – Southwest anchor

So, anchors are used as reference points for positioning nodes and drawing between nodes. Anchors can be used as coordinates. That explains why node names and anchors are written in parentheses like coordinates.

Since the anchors of node shapes are very useful, we will look at them in detail in the next section.

Using shapes and anchors

While rectangle and circle node shapes are available by default, others require loading the `shapes` package, as we did in the previous section.

We will explore many of them now.

A rectangle shape

A `rectangle` node has anchors in all compass directions, as we can see here, with a node named `(n)`:

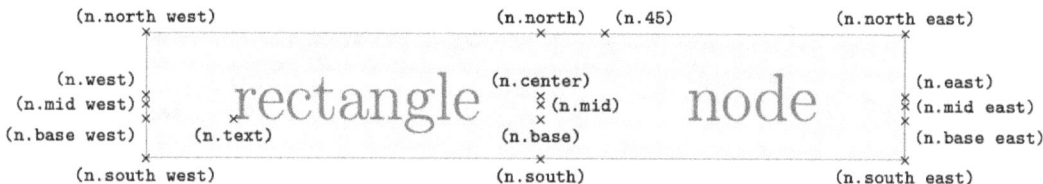

Figure 3.6 – Rectangle shape with anchors

In addition to these, we have a few more anchors available:

- `center`: The middle of the node, which is the default anchor.
- `base`: At the baseline of the node text and centered horizontally. It is helpful for the vertical alignment of text nodes. The `base west` and `base east` anchors are at the baseline height and on the west and east sides, respectively.
- `text`: At the left of the text baseline.
- `mid`: At half-height of the lower x and centered horizontally. It is also helpful for vertically aligning nodes with text that may have different heights and depths. Also, here, `mid east` and `mid west` are at the `mid` height and west and east sides, respectively.
- `(n.45)`: This is an anchor on the border with an angle of 45 degrees to the center. You can use any angle in degrees counterclockwise; negative values are also accepted. For example, `(n.90)` would be the same as the `north` anchor, and `(n.180)` would be the same as the `west` anchor.

The angle anchors are handy for manual adjustment when the standard anchors are not a good fit.

The circle and ellipse shapes

The `circle` shape provides the same anchors as the `rectangle` shape, adapted to a circle:

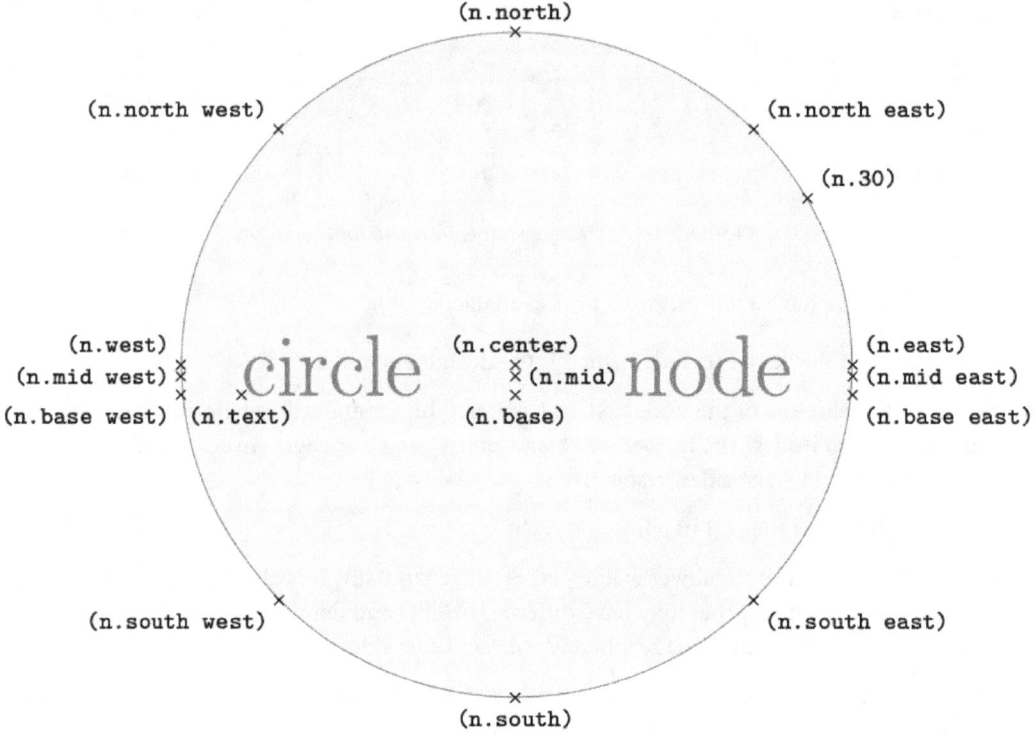

Figure 3.7 – Circle shape with anchors

The angle anchors look more intuitive here, such as `(n.30)`, which is at the circular border of the `(n)` node at 30 degrees counterclockwise.

The `ellipse` shape provides the very same anchors.

The coordinate shape

In a drawing, we may have specific coordinates, that may be, for example, starting points for lines or arrows or reference points for placement of other nodes. We can give coordinates a name to refer to them by name instead of by numerical values.

For example, this names the (2,0) coordinate begin and the (4,2) coordinate end:

```
\coordinate (begin) at (2,0);
\coordinate (end)   at (4,2);
```

From now on, we can use those names as coordinates in our drawing, like this, drawing a line between the coordinates:

```
\draw (begin) -- (end);
```

Using named coordinates makes complex drawings more readable, and it separates numerical values from the semantics of our drawing commands.

Technically, such a named coordinate is a node. We could also have defined the `begin` coordinate that way:

```
\node[shape=coordinate] (begin) at (2,0) {};
```

This has the same effect. So, a coordinate is a node with empty text and the coordinate shape, meaning it has zero width and height values. It has the same anchor names as a default rectangle node, but of course, all anchors are equal here, so you don't need to specify any anchor.

Our use case here is to get names for numerical values, helping us structure a drawing. In contrast to a node with shape, dimensions, and text, we can consider such a named coordinate a geometrical point.

More shapes

There are many more highly customizable shapes; see the *Further reading* section, at the end of this chapter.

Here is a quick example collection of what the shapes are called and how they look:

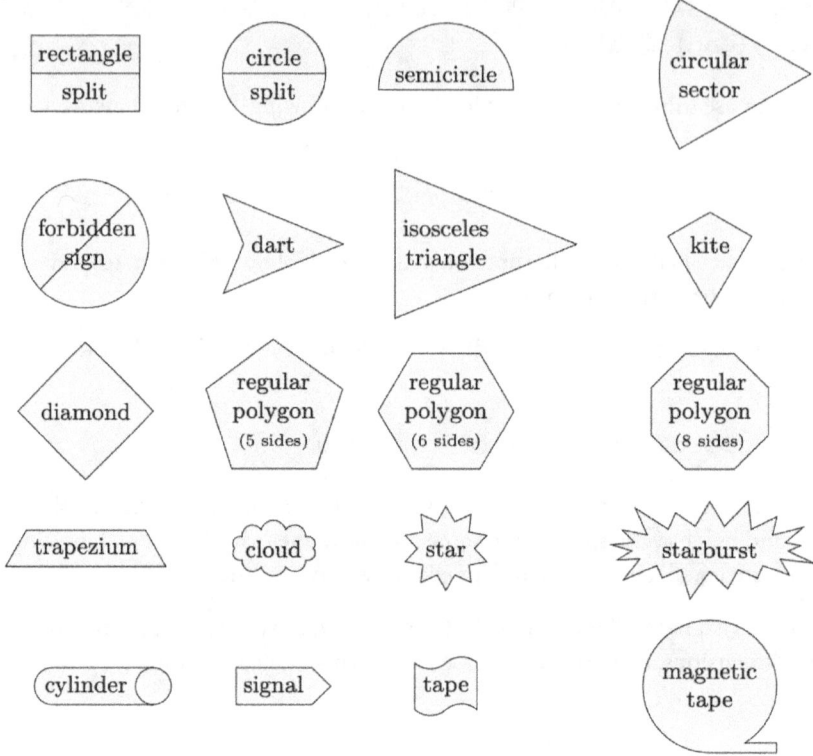

Figure 3.8 – Various node shapes

Many shapes provide particular options, such as the number of puffs in a cloud, the number of parts in a split rectangle, aspect ratio, angles, and of course, the standard options for color, filling, rotation, line width, and many more.

Once nodes and anchors are understood, it's often not much more complicated than selecting the desired shape, using the comprehensive manual to choose from the available design options for shapes, selecting colors, and then doing some fine-tuning on dimensions.

Let's play with shapes to get into a routine with TikZ node shapes. Apart from libraries, other packages use TikZ and build on it. One is the `tikzpeople` package, which provides shapes of people. It was originally intended to depict the usage of cryptographic protocols between parties. Now it's our turn with it. We can load it using the following:

```
\usepackage{tikzpeople}
```

At https://texdoc.org/pkg/tikzpeople, we can read the documentation, and we find that there's a `graduate` shape and a `monitor` option. All of the following, we do in the usual `\begin{tikzpicture}` ... `\end{tikzpicture}` environment. Let's draw a node that we will call `student` in front of a monitor.

```
\node (student) [graduate, monitor, minimum size=2cm] {};
```

Then we try the `starburst` shape from *Figure 3.8*. In the TikZ manual, *Part V: Libraries, 71: Shape Library, in the Symbol Shapes subsection*, we find options for the shape and customize it to get an awesome error message node. `inner sep`, for example, is the space between the node text and the node border. We will see it again in the next section, *Spacing within and around nodes*.

We position the `starburst` node at the `student` node at 45 degrees:

```
\node at (student.45) [starburst, draw=red, fill=yellow,
    starburst point height=0.4cm, line width=1pt,
    font=\ttfamily\scriptsize, inner sep=1.5pt] {error};
```

Next, we use the `cloud callout` shape from *Figure 3.8* to depict a student's thought bubble. We find the customization options in the TikZ manual, in the *Callout Shapes* section of the aforementioned chapter. We choose nice-looking options, such as an aspect ratio of 3 and ball shading. Specifically, we define the anchor of the callout node to be `pointer`, which is the smallest bubble here. And we position the node pointer anchor right at an angle of 130 degrees of the student:

```
\node at (student.130) [cloud callout, cloud puffs=13,
    aspect=3, anchor=pointer, shading=ball,
    ball color=darkgray, text=white, font=\bfseries]
    {My thesis...!};
```

Now compile the document with the picture, and let's look at our three nodes:

Figure 3.9 – Positioning node shapes

Nobody can keep all shape and node options in mind. That's why it's widespread to use the TikZ manual as a reference all the time, and then the challenge is just to read and apply options.

At `https://tikz.net/tag/shapes`, you can find a collection of TikZ examples specifically for using shapes.

We can see that default node distances and spacing are pretty good. Still, we may want to customize them, so this will be our next topic.

Spacing within and around nodes

We saw that rectangular node borders just fit nicely around the text. To understand how a circular node border fits around the node text: imagine a rectangle node for this text, and then the circle node border circumscribes that rectangle.

You can set a node option called `inner sep` to get more or less distance between the node text and border. To get more spacing around the border so the anchors are farther away, you can set an optional value called `outer sep`. It's written in the following way:

```
\node[draw,rectangle,inner sep=1cm,outer sep=1cm] {n};
```

Spacing within and around nodes   35

It is better to see it in a picture, so take a look with a default spacing node next to the n node:

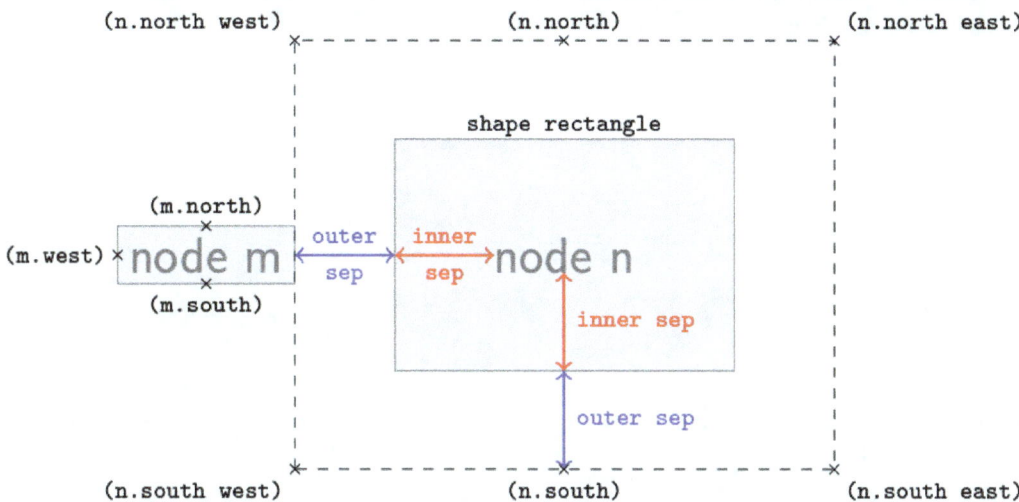

Figure 3.10 – Spacing within and around a node

We can set horizontal (x) and vertical (y) distance separately; they are called xsep and ysep. With example values of 1cm and 0.5cm, the code changes to the following:

```
\node[draw,rectangle, inner xsep=1cm,inner ysep=0.5cm,
   outer xsep=1cm,outer ysep=0.5cm] {n};
```

In *Figure 3.11*, we can see how the distances change. We have much more horizontal spacing and less vertical spacing:

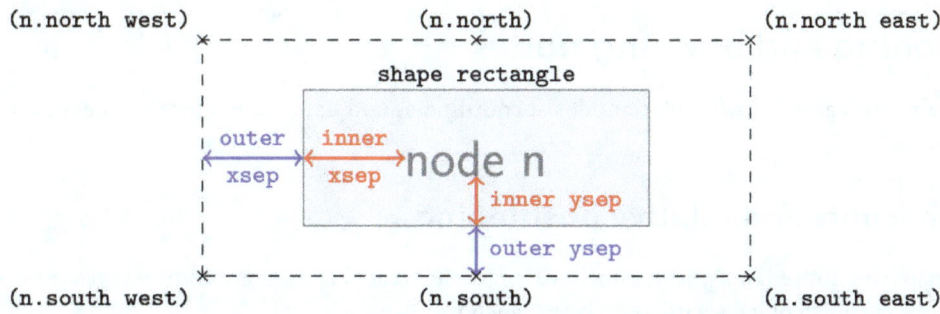

Figure 3.11 – Different horizontal and vertical spacing

You can insert spacing in the same way with circle nodes. However, remember that the circular border is actually around the imagined rectangle node, and the `inner sep` value determines the rectangle's internal spacing:

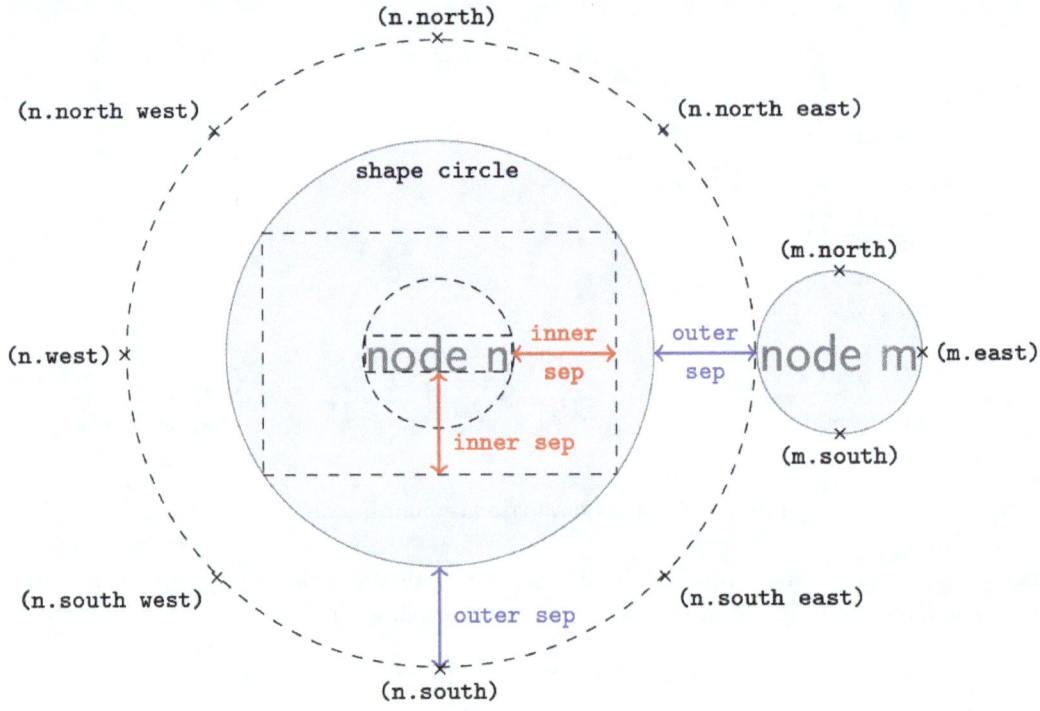

Figure 3.12 – Spacing within and around a circle node

Now, let's dig deeper into the positioning of nodes.

Positioning and aligning nodes

We have already learned how to place nodes at coordinates and use anchors for that. Let's explore more options.

Using anchors and relative positioning

First, perhaps you noticed that positioning based on anchors can feel counterintuitive: to place a node above an object (north of it), we use the `south` anchor.

For example, here we draw a node above a circle:

```
\draw circle [fill, radius=2pt] node [anchor=south] {text};
```

The output of that command is shown in the following picture:

Figure 3.13 – A node above a circle

For more intuitive positioning, TikZ offers other statements. We can write the same line in this way:

```
\draw circle [fill, radius=2pt] node [above] {text};
```

That gives the same output as in *Figure 3.13*, and feels more natural.

In that spirit, these are the new relative positioning options:

- `above`: Similar to `anchor=south`
- `below`: Similar to `anchor=north`
- `left`: Similar to `anchor=east`
- `right`: Similar to `anchor=west`
- `above left`: Similar to `anchor=south east`
- `above right`: Similar to `anchor=south west`
- `below left`: Similar to `anchor=north east`
- `below right`: Similar to `anchor=north west`
- `base left`: Similar to `anchor=base east`
- `base right`: Similar to `anchor=base west`

This can also be used for relative positioning between nodes. For this, we will use the `positioning` library. In our document, we have to add this to our preamble:

```
\usetikzlibrary{positioning}
```

Now, let's say we draw a node called `TikZ` by doing the following:

```
\node [draw] (TikZ) {TikZ};
```

Then, we can draw another node 0.1 cm right of it, as follows:

```
\node [draw, right = 0.1cm of TikZ] {PDF};
```

The output is shown in the following picture:

Figure 3.14 – A node to the right of another node

The other directions work similarly. Note that we can specify above/below and left/right offsets separately. That's done with the and keyword, such as here:

```
\node [draw, above right = -0.25cm and 0.1cm of TikZ]
  {PDF};
```

This would give us the PDF node above right, but shifted -0.25 cm vertically:

Figure 3.15 – A node above and to the right of another node

This gives us an easy way of arranging nodes without too much headache caused by coordinates.

Placing nodes along a line

With a single \draw or \path command, we can place several nodes along the path between two coordinates. The primary purpose is to set text on the one hand over or under a line, with the same options we had in the previous section. On the other hand, we may want to place nodes at the start, the end, or the middle of a line.

Consider a line between the (0,0) and (4,0) coordinates:

```
\draw (0,0) -- (4,0);
```

We can insert a node by using node [pos=value] with a value between 0 and 1. 0 means at the start of the line, 1 means at the end of the line, and any value in between means at the corresponding fraction of the line. So, pos=0.5 means at the middle of the line.

Let's see it with a picture. We use the following code:

```
\draw (0,0) --
    node [above,    pos=0]     {0}
    node [above,    pos=0.5]   {0.5}
```

```
node [above,    pos=1]     {1}
node [below, pos=0.25] {0.25}
node [below, pos=0.75] {0.75}
(4,0);
```

This gives us the following picture:

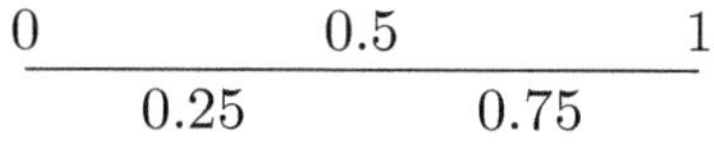

Figure 3.16 – Nodes along a line

There are also predefined options for certain positions:

- `at start`: Like pos=0. `node [at start] {...}`, it places the node at the beginning of the line
- `very near start`: Like pos=0.125, very close to the start
- `near start`: Like pos=0.25, close to the start
- `midway`: Like pos=0.5, at the middle of the line
- `near end`: Like pos=0.75, close to the end
- `very near end`: Like pos=0.875, very close to the end
- `at end`: Like pos=1, at the end of the line

This positioning also works along paths that are bent or curved in any way.

Aligning nodes at the text baseline

In *Figure 3.14*, we were lucky that the texts in the nodes had the same size. That's not always the case, so let's prepare for the other situation when the node texts are of different heights, such as capital letters, small letters, or letters with descenders.

Let's put five nodes next to each other to form **Epic.** as a phrase. Each node will contain a letter or dot. We set `inner sep` to 0 points so that the nodes will be next to each other.

The straightforward code is as follows:

```
\begin{tikzpicture}[every node/.style = {inner sep=0pt}]
  \node (E)      {E};
  \node (p) [right = 0pt of E] {p};
  \node (i) [right = 0pt of p] {i};
  \node (c) [right = 0pt of i] {c};
```

```
    \node (.) [right = 0pt of c] {.};
\end{tikzpicture}
```

Now we may expect to get **Epic.** as text. Compile, and see how it looks for now:

$$\text{Epic.}$$

Figure 3.17 – An epic misalignment

It looks like an epic failure. Let's fix it. We can use the `base` alignment briefly mentioned in the alignment options earlier in this section. Change each `right` to `base right`:

```
\node (E) {E};
\node (p) [base right = 0pt of E] {p};
\node (i) [base right = 0pt of p] {i};
\node (c) [base right = 0pt of i] {c};
\node (.) [base right = 0pt of c] {.};
```

Compile again, and now we get the following output:

$$\text{Epic.}$$

Figure 3.18 – Epic base alignment

That's what we want! This example was to stress the importance of base alignment.

Aligning whole pictures at a node text baseline

The base anchor of a node can even be used to align the complete TikZ picture to the surrounding text. That's beneficial if we use small TikZ images inline in text paragraphs or place two TikZ pictures with text content next to each other.

Let's take a simple example. We will use a circled number 1 as the label for the first topic:

```
\begin{tikzpicture}
  \node[circle, draw, inner sep=2pt] (label) {1};
\end{tikzpicture}
This is the first topic.
```

As it is, we get the following as the output:

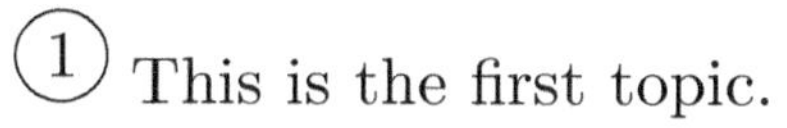

① This is the first topic.

Figure 3.19 – Default picture alignment

While bottom alignment may be okay for image nodes, that's not our choice for text nodes. In our case, we shall use the text node anchor for aligning, that is, `label.base`. Set the following as our baseline for the whole picture:

```
\begin{tikzpicture}[baseline=(label.base)]
```

With this change, the output becomes nicely aligned:

① This is the first topic.

Figure 3.20 – TikZ picture baseline alignment

Remember the short syntax; we can use the `\tikz` command instead of a `\begin{tikzpicture}` … `\end{tikzpicture}` environment. That may be handy for concise TikZ picture definitions. Our code for *Figure 3.20* would become shorter:

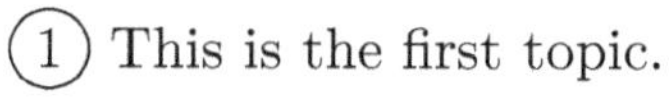

```
\tikz[baseline=(label.base)]{\node[circle, draw,
   inner sep=2pt] (label) {1};}
```

For us LaTeX users, it's natural to define a macro for circled numbers. Let's do this small exercise here, similar to what I wrote online at `https://tex.stackexchange.com/a/7045`. We use `\DeclareRobustCommand`, so our macro is not breaking at certain places, such as captions, labels, or headings; simply said, that's what LaTeX users call **robust**. Its syntax is similar to `\newcommand`. We support one argument for the number in the circle. So, a macro definition for our previous TikZ command can be made like this:

```
\DeclareRobustCommand{\circled}[1]{%
   \tikz[baseline=(label.base)]{\node[circle, draw,
   inner sep=2pt] (label) {#1};}}
```

Now that we have defined the command, *Figure 3.20* can be more easily produced by the following line:

```
\circled{1} This is the first topic.
```

This is much more manageable, and we can continue with `\circled{2}`, `\circled{3}`, and so on. We can even use it in the LaTeX `enumerate` environments to show further how we can integrate LaTeX environments and macros with TikZ.

We load the `enumitem` package that lets us customize the `enumerate` environments:

```
\usepackage{enumitem}
```

We use our command definition and add a bit of color to look fancier:

```
\DeclareRobustCommand{\circled}[1]{%
  \tikz[baseline=(label.base)]{\node[circle,
    white, shading=ball, inner sep=2pt] (label) {#1};}}
```

We create an `enumerate` environment, where we declare that the label should be circled with Arabic numbers:

```
\begin{enumerate}[label=\circled{\arabic*}]
  \item First item
  \item Second item
  \item Third item
  \item Fourth item
\end{enumerate}
```

When you compile this, the automatic item numbering is now fancy with well-aligned TikZ images:

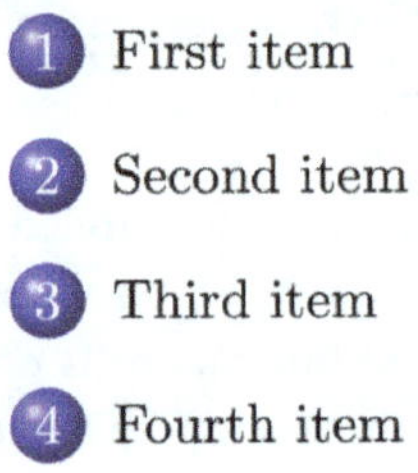

① First item

② Second item

③ Third item

④ Fourth item

Figure 3.21 – An enumerate environment with fancy TikZ numbers

Note that the default shading ball color is blue. We will discuss shading and colors in *Chapter 7*, *Filling, Clipping, and Shading*.

Now we have talked a lot about positioning and aligning nodes, there's another quick way to place nodes next to other nodes: using labels. Let's look at this next.

Adding labels and pins

We can add labels to nodes with a handy syntax that looks like this:

```
\node[label=direction:text] at (coordinate) {text};
```

Note that if we don't specify a `coordinate` value, the node will be at the current position in the path. Paths begin at the origin (0,0) by default if no `coordinate` value is specified. Knowing this, we will omit the `coordinate` value in the following examples, so our nodes will be at (0,0).

Again, it's good to see it in a picture. Let's have a ball node with labels, where every label is scaled down by two.

For this, we will first have a brief look at the style syntax, as it's already convenient here. Until now, we set the `key=value` pairs as options for nodes or other elements. To not repeat ourselves, we can set these options for all elements in a drawing by using a single option on the `tikzpicture` environment:

```
\begin{tikzpicture}[every node/.style={key=value}]
```

The dot is part of the syntax we thoroughly discuss in *Chapter 5, Using Styles and Pics*. For now, we can also apply that to labels. Let's go:

```
\begin{tikzpicture}[every label/.style = {scale=0.5}]
  \node[
    label = above:Graphics,
    label = left:Design,
    label = below:Typography,
    label = right:Coding,
    circle, shading=ball, ball color=blue!60,
      text=white] {TikZ};
\end{tikzpicture}
```

This gives us the following picture:

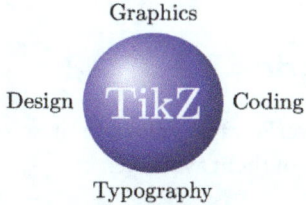

Figure 3.22 – A node with labels

This works with angles as well, so `label = {90:Graphics}` would have the same effect as `label = {above:Graphics}`.

If we replace `label` with `pin`, we get labels connected to the origin node by lines:

```
\begin{tikzpicture}[every pin/.style = {scale=0.5}]
  \node[
    pin = above:Graphics,
    pin = left:Design,
    pin = below:Typography,
    pin = right:Coding,
    circle, shading=ball, ball color=blue!60,
      text=white] {TikZ};
\end{tikzpicture}
```

With `pin`, the output becomes the following:

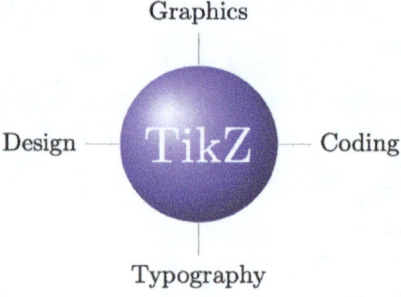

Figure 3.23 – A node with pinned labels

That's quite handy for a single node with some annotations.

The content of a node doesn't have to be pure text; we can even insert images. Let's try this in the next section.

Putting images into nodes

We all know about fancy Visio and PowerPoint diagrams. In these, we have fancy node shapes, which are called **icons** or **stencils**, with many of them available.

While TikZ gives us a library of various shapes that we can customize, we can even use arbitrary images as nodes that we combine with a shape.

I'm working as a network engineer and producing complex network diagrams in my field of work. So, I will describe my approach.

Renowned hardware manufacturers, such as Cisco and Hewlett Packard often provide icon and stencil libraries for use with Visio, PowerPoint, Inkscape, or any drawing program. We can use the same in TikZ. So, we can go to a vendor download page, such as https://www.cisco.com/c/en/us/about/brand-center/network-topology-icons.html. There we can find image collections in various formats, such as .vss for Visio, .pptx for PowerPoint, .jpg for general use, and .eps in **Encapsulated PostScript** format (**EPS**).

The best choice here is EPS because, in contrast to JPG files, EPS files are *scalable*. That means we can use them in small or large sizes without quality loss. LaTeX supports EPS. However, with pdfLaTeX and for producing PDFs in a straightforward way, we better convert them into the PDF format, with the same benefit of scalability.

The **epstopdf** tool transforms an EPS image into a PDF format with the same size. We choose a router image and a switch image from the manufacturer's collection. At the command line, the epstopdf router.pdf command transforms a file named router.eps into router.pdf file. We can do the same to get a switch.pdf file. We can now use that together with the \includegraphics command from the graphicx package that's already loaded by TikZ implicitly.

The following line gives us a router node with a size of 2 cm and 0 inner sep to avoid unnecessary spacing:

```
\node (router) [inner sep=0pt]
  {\includegraphics[width=2cm]{router.pdf}};
```

This gives us a switch node right next to the router node:

```
\node (switch) [inner sep=0pt, right = of router]
  {\includegraphics[width=2cm]{switch.pdf}};
```

For our first network engineering drawing, we draw a double line to depict the cabling between the router and the switch:

```
\draw[double] (router) - (switch);
```

Those three TikZ commands already give us a pretty good start for a network drawing:

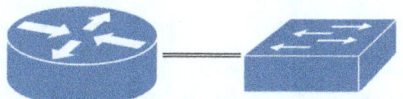

Figure 3.24 – Images in nodes

Since we did not specify a node shape, the default rectangle shape is used. That's a good fit in our case, and we can use the usual `rectangle` shape anchors for placing other lines and labels there. Depending on the image, you can, of course, choose another node shape, such as an `ellipse` or `circle` shape.

In technical drawings, we will have to add a lot of labels, which are other text nodes. Here, let's have the following:

- A `TenGig` label above the line documents a 10-gigabit connection
- A `1` label below the line at its beginning states the router port `1`
- A `24` label below the line at its end determines the switch port `24`

The following line extends the previous \draw command and adds nodes to the lines as desired:

```
\draw[double] (router) --
    node [above, font=\scriptsize] {TenGig}
    node [font=\tiny, inner xsep=0pt,
      below right, at start] {1}
    node [font=\tiny, inner xsep=0pt,
      below left, at end] {24}
  (switch);
```

Compile, and the drawing now has text labels above and below the connection line:

Figure 3.25 – Images in nodes with connection and labels

That wasn't too hard – a quick drawing with a few commands based on external images. When we use more symbol nodes, more labels, and different kinds of connections, we don't have to repeat commands for font size, `inner xsep`, or `\includegraphics` with filenames all the time: we can define styles for our own node types. In *Chapter 5, Using Styles and Pics*, we will learn about this.

Summary

Now, we have learned about the concept of nodes in TikZ. Using nodes, we can now add text to our drawings with complete control over its placement and alignment. Furthermore, we learned how to put shapes around our text and how to use external images within TikZ nodes.

Nodes will be the building blocks of your TikZ diagrams and drawings, so mastering this chapter was an important step forward.

In the next chapter, we will see how to connect nodes with edges and arrows.

Further reading

The TikZ manual explains nodes in *Part III, Section 17, Nodes and Edges*. You can open the manual by typing `texdoc tikz` at the command prompt or online at `https://texdoc.org/pkg/tikz` as a PDF document. You can read that chapter in an online HTML version of the manual at `https://tikz.dev/tikz-shapes`.

Furthermore, the TikZ manual has a comprehensive reference of shapes in *Part V, Section 71, Shape Library*, also online at `https://tikz.dev/library-shapes`. It shows all shapes with anchors and many customizations.

The `tikzpeople` package documentation is available at `https://texdoc.org/pkg/tikzpeople`.

The home page for the `epstopdf` tool is `https://tug.org/epstopdf/`; there, you can find links to download it and documentation.

4
Drawing Edges and Arrows

In the previous chapter, we learned how to produce shapes with text called **nodes**. In this chapter, you will learn how to draw lines with text, called **edges**, to complete your knowledge about placing text in diagrams.

We will also discuss customizing edges, texts, and arrows.

In this chapter, we will cover the following key topics:

- Connecting nodes by edges
- Adding text to edges
- Diving deeper into edge options
- Drawing arrows
- Using the `to` operation

By the end of this chapter, you will be able to draw colorful diagrams with text, arrows, and edge labels.

Technical requirements

You need to have a LaTeX installation, including TikZ, or you can compile the example code online with Overleaf or at `https://tikz.org/chapter-04`.

The code is also available on GitHub at `https://github.com/PacktPublishing/LaTeX-graphics-with-TikZ/tree/main/04-drawing-edges-and-arrows`.

From time to time, you will see code snippets for an explanation. You can take the corresponding complete source code from TikZ.org or GitHub if you want to run such a snippet.

In this chapter, we will use the following TikZ libraries: `positioning`, `quote`, and `arrows.meta`. Furthermore, we will use the `topaths` library, which is loaded automatically by TikZ, so we don't have to load it ourselves.

Connecting nodes by edges

In the previous chapter, we used lines with arrows to create the small diagram in *Figure 3.3*. In more complex diagrams, text labels on such lines and arrows are often desired. Such "smart" lines connecting nodes are called edges.

We will start with the first small example. We aim to illustrate the compiling process from a LaTeX `.tex` file to a PDF file. We will also add some colors to get more familiar with styling nodes.

1. Start with this small LaTeX document, which loads TikZ and the `positioning` library and contains an empty `tikzpicture` environment for now:

   ```
   \documentclass[border=10pt]{standalone}
   \usepackage{tikz}
   \usetikzlibrary{positioning}
   \begin{document}
   \begin{tikzpicture}
   \end{tikzpicture}
   \end{document}
   ```

2. In the `tikzpicture` environment, create a node, which we call `tex`, filled with orange color and white text:

   ```
   \node (tex) [fill=orange, text=white] {TEX};
   ```

3. Put a second node on the right of the `tex` node, call it `pdf`, and fill it with a color that reminds us of the Adobe Acrobat Reader logo:

   ```
   \node (pdf) [fill={rgb:red,244;green,15;blue,2},
     text=white, right=of tex] {PDF};
   ```

4. Now, we draw our first edge with an arrow tip from the `tex` node to the `pdf` node:

   ```
   \draw (tex) edge[->] (pdf);
   ```

5. Compile the document, which will present the following picture:

Figure 4.1 – A basic edge connecting two nodes

Step 1 is typically the code we start from throughout this book. While we chose a simple orange node color in *Step 2*, in *Step 3*, we used the RGB value (244,15,2), representing Adobe's colors, the PDF format creator.

In *Step 4*, we saw our first edge. The syntax typically used is as follows:

```
(node1) edge[options] (node2)
```

That can be used with a \draw command, as we did here, or with the \path command.

Let's look at how we can put text on the edges.

Adding text to edges

In diagrams, we often see that apart from text in diagram nodes, we can have text on the connecting lines or arrows. That's an essential feature of TikZ's edge operation.

Let's continue our example from the previous section and add a text label to the edge. It will read pdflatex in a tiny typewriter font printed above the edge. This label is itself a node, so we insert this right after the edge:

```
node[font=\tiny\ttfamily, above] {pdflatex}
```

The full command becomes as follows:

```
\draw (tex) edge[->]
  node[font=\tiny\ttfamily, above] {pdflatex} (pdf);
```

Compile, and you get this picture:

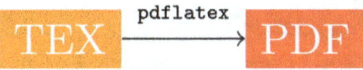

Figure 4.2 – An edge with a text label

Admittedly, this is a pretty verbose syntax. Luckily, TikZ provides a shorter way; this is called the **quotes syntax** because you can add edge label texts by enclosing the text in quotes as an option to the edge. That's basically edge["text"]. We can set style options to the quoted text, for example, edge["text" red], to get a red text. If you have several options, enclose them in curly braces. Otherwise, TikZ will not know whether an option after a comma is an option for the quoted text or an option for the edge. Applying this, our edge drawing command becomes the following:

```
\draw (tex)
  edge["pdflatex" {font=\ttfamily\tiny,above},->] (pdf);
```

With this command, we get the same output as in *Figure 4.2*.

In the previous chapter, we introduced the `style` syntax. Let's take the same approach here. We can define the following options:

- **A style for all nodes**: We choose a white text color and a minimum width of 1.1 cm, so the nodes have the same width even if they have less text in them
- **A style for all edges**: In our case, these shall be drawn as arrows
- **A style for all edge quotes**: We want the automatic placement of text next to the edges, in small typewriter font, and in a black text color

We can write these definitions as options for the `tikzpicture` environment:

```
\begin{tikzpicture}[
  every node/.style = {text=white, minimum width=1.1cm},
  every edge/.style = {draw,->},
  every edge quotes/.style = {text=black,
    auto, font=\ttfamily\tiny, inner sep=1pt}]
]
```

We benefit from such a definition when we have more nodes and edges. So, let's create four nodes now:

```
\node (tex) [fill=orange] {TEX};
\node (pdf) [fill={rgb:red,244;green,15;blue,2},
  right=of tex] {PDF};
\node (dvi) [fill=blue, above=of tex] {DVI};
\node (ps)  [fill=black!60, above=of pdf] {PS};
```

Without any edges, this gives the following picture:

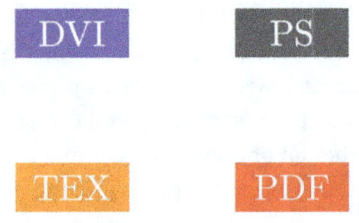

Figure 4.3 – Four nodes

We did not have to repeat `text=white` for each node. Now, we can add many edges. Insert these commands after the nodes:

```
\draw (tex) edge ["pdflatex"]  (pdf)
      (tex) edge ["latex"]     (dvi)
```

```
              (dvi)  edge["dvips"]      (ps)
              (dvi)  edge["dvipdfmx"]   (pdf)
              (ps)   edge["ps2pdf"]     (pdf);
```

Compile again, and the picture now becomes what is shown in *Figure 4.4*:

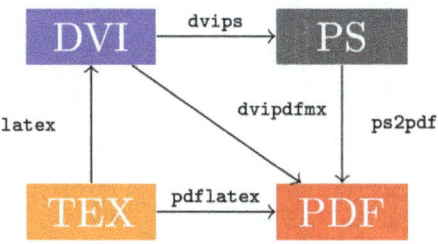

Figure 4.4 – Many edges with text

We did not have to repeat the style commands for the edge texts. Thanks to the general style definition, the `quotes` syntax for edge texts is concise, clear, and readable.

When we use a single `\draw` command for all edges, it's even cleaner because we can omit the start node of an edge if it's the same start node that the previous edge had:

```
    \draw (tex)  edge["pdflatex"]   (pdf)
                 edge["latex"]      (dvi)
          (dvi)  edge["dvips"]      (ps)
                 edge["dvipdfmx"]   (pdf)
          (ps)   edge["ps2pdf"]     (pdf);
```

Styles save us from much work and result in cleaner code. Even better, when we modify the style, it influences all elements with that style. For example, let's add the `sloped` option to our quotes style so the edge texts are rotated to follow the path:

```
    every edge quotes/.style = {text=black, auto,
      font = \tiny\ttfamily, inner sep=1pt, sloped}]
```

By adding that single option, all edge texts are now written nicely along the edges:

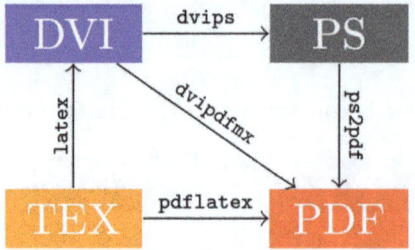

Figure 4.5 – Sloped text on edges

We will discuss styles further in *Chapter 5, Using Styles and Pics*.

The auto option can be modified: by default, it ensures that the text is written on the left side of an edge, which is the same as setting `auto=left`. We can decide that the text should go to the right side of an edge by choosing `auto=right`. The effect on our previous example would be as follows:

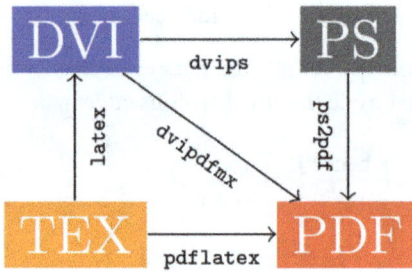

Figure 4.6 – Sloped text with the auto=right option

Note that with nodes on edges, we can apply the same positioning options we saw in *Chapter 3, Drawing and Positioning Nodes*, in the *Positioning and aligning nodes* section.

We can even have curvy edges; let's look at more straight and curvy options next.

Diving deeper into edge options

Every `edge` operation builds its own path; that's why `edge` supports general path options. Furthermore, an edge can have connection-specific options, such as defining a straight line or a curve.

So, we can divide edge styling options into the following:

- Path options, including general TikZ options
- Connection options

Let's look at them separately.

Path options

Remember, a path is a series of lines or curves with nodes and other elements. All options that you can set to a path, you can set to an edge, such as line style options.

These options can be as follows, using example values for clarity instead of syntax grammar formality:

- `color=red`: This draws the edge in a red color
- `ultra thin, very thin, thin, semi thick, very thick, ultra thick`: These options define the thickness of the edge line
- `line width=3pt`: This denotes that the edge shall have a thickness of 3 pt
- `loosely dotted, dotted, densely dotted`: These define that the edge has a dotted pattern with more or less density
- `loosely dashed, dashed, densely dashed`: These define that the edge has a dashed pattern with more or less density
- `loosely dash dot, dash dot, densely dash dot`: These define that the edge has dashed and dotted patterns with different densities
- `loosely dash dot dot, dash dot dot, densely dash dot dot`: These again define different density patterns; this time it's a dash with two following dots
- `line cap=round, line cap=rect, line cap=butt`: These define edge lines that are end rounded, or with rectangle closing, or just cut at the end
- `double=yellow`: This draws a double-edge line with the inner color yellow, while the border color is the same as what is set by the `color` option
- `double distance=2pt`: This draws a double edge link above, with a spacing of 2 pt between the border lines
- `transform canvas={yshift=5pt}`: This shifts an edge up in the y direction by 5 pt; use or add `xshift` for moving in the x direction

We will see examples for the last three options in the next chapter.

Connection options

Apart from general path options, some options define the properties of a connection itself, specifically if it is a curve rather than a line:

- `out=45`: The edge curve will leave the start coordinate at an angle of 45 degrees.
- `in=90`: The edge curve will arrive at the target coordinate at an angle of 90 degrees.

- `relative=true, relative=false`: If `true`, in and out angles are seen as relative to a direct linear connection between the start and target node (by default), or are seen as absolute if `false`, that is, relative to the paper.
- `bend left=30, bend right=30`: The edge will bend by 30 degrees to the left or right. *Figure 5.3* shows an example of this.
- `looseness=0.5, looseness=1.5`: The bent edge will be tighter (0.5) or looser (1.5). The default value is 1, meaning that the curve behaves like a circle or like an arc when in and out values allow, but the looseness factor makes it tighter or less tight.
- `in looseness, out looseness`: You can choose different values for `looseness` regarding the `in` and `out` connections.
- `min distance, max distance`: These restrict the preceding calculated bending to a minimum or maximum distance.
- `in min distance, in max distance, out min distance, out max distance`: Like above, but for the `in` and `out` directions, respectively.
- `distance`: No computed value; the edge curve will go that far in bending distance.
- `loop`: The edge connects to itself, no matter what coordinate follows; you can specify the `out`, `in`, and `looseness` values as you desire.

We will use some of these options in *Chapter 12, Drawing Curves*, where you can find examples. In addition, we will use Bezier curve connections there.

Arrow tips are path options, and the next section is dedicated to those.

Drawing arrows

We already saw how to add basic arrow tips to lines and edges: we added the `->` option to get a right arrow tip. However, the default arrow tips look a bit small and thin. Let's see how to modify them.

The general syntax is `\draw[starttip-endtip]` or `edge[start arrow-endarrow]`. Previously, we had just `>` as the end tip and no start tip.

Quick examples are provided in the following list:

- `->`: Right arrow, `<-`: Left arrow
- `<->`: Left and right arrow
- `->>`: Double right arrow, `<<-`: Double left arrow
- `-Triangle`: Triangle-shaped arrow
- `-Stealth`: Stealth-plane-shaped arrow
- `-LaTeX`: Black triangle arrow with slightly bent sides, like `\vec` in LaTeX picture mode

You can combine one or several tops at an arrow's start and end.

When you use arrows, always load the `arrows.meta` package:

```
\usetikzlibrary{arrows.meta}
```

The old TikZ `arrows` library should not be used anymore. It still exists; that's why the new library has a different name, where `meta` was added to indicate a similarity to meta-fonts. When we change the size of an arrow, it's not simply scaled, but TikZ does a complex size transformation, so arrow tips grow slower than the line width to have a good appearance. This library offers many ways of customizing width, height, shape, filling, and thickness.

Let's look at available arrow tips and customization next.

Mathematical arrow tips

These arrow tips are drawn like the classic arrow tips from standard LaTeX mathematical fonts:

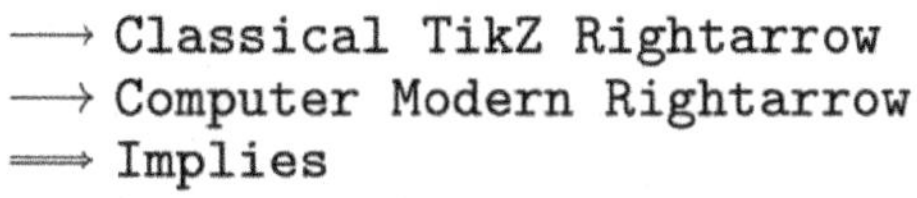

⟶ Classical TikZ Rightarrow
⟶ Computer Modern Rightarrow
⟹ Implies

Figure 4.7 – Mathematical arrows

The `Implies` arrow is for double lines or edges, such as in this code:

```
\draw (node1) edge[-Implies, double] (node2);
```

To is a shortcut for `Computer Modern Rightarrow`.

Barbed arrow tips

Like the mathematical arrow tips, barbed arrow tips are like lines on classic arrows, not filled tips like a triangle:

⟶ Arc Barb
⟶ Bar
⟶ Bracket
⟶ Hooks
⟶ Parenthesis
⟶ Straight Barb
⟶ Tee Barb

Figure 4.8 – Barbed arrows

The dimensions of the tips can be customized; we will look at them after we look at the tip types.

Geometric arrow tips

These arrow tips are closed shapes. They can be filled with black or another color, or they can be open:

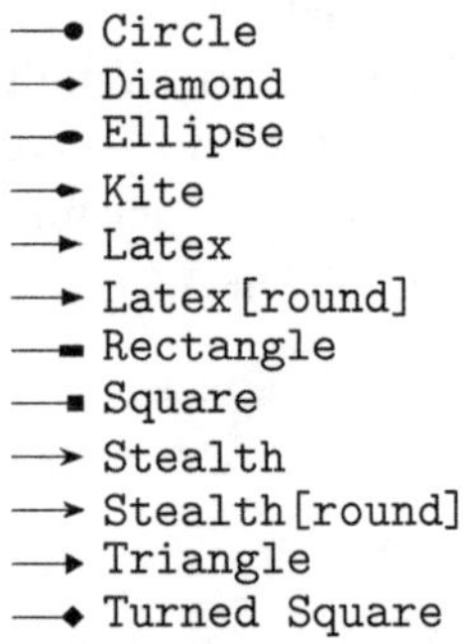

```
→• Circle
→◆ Diamond
→● Ellipse
→◆ Kite
→► Latex
→► Latex[round]
→■ Rectangle
→■ Square
→► Stealth
→► Stealth[round]
→► Triangle
→◆ Turned Square
```

Figure 4.9 – Geometric arrows

Here, we have open types:

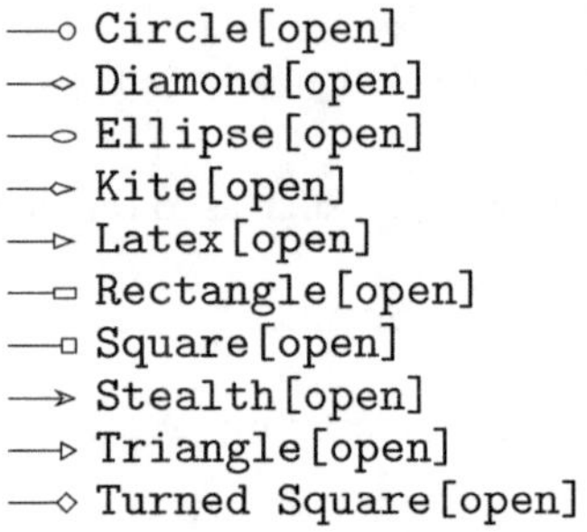

```
→○ Circle[open]
→◇ Diamond[open]
→○ Ellipse[open]
→◇ Kite[open]
→▷ Latex[open]
→□ Rectangle[open]
→□ Square[open]
→▷ Stealth[open]
→▷ Triangle[open]
→◇ Turned Square[open]
```

Figure 4.10 – Open geometric arrows

Also, here, the dimensions of the tips can be customized, so let's now turn to this.

Customizing arrow tips

These are the customization options:

- `length`: Defines the length of the arrow tip in the direction of the line or edge
- `width`: Defines how wide the arrow is; you can consider it also like its height
- `scale`: This is a scaling factor
- `scale length` and `scale width`: These are like `scale` but only for length or width, respectively

- `slant`: A factor that slants the tip
- `reversed`: Draws the tip in reversed direction
- `harpoon`: Draws only the left (or call it upper) half of the arrow tip
- `swap`: Flips the arrow along the line; this is useful if you want to have a harpoon tip on the other side so you can combine them
- `color`: Draws the arrow tip with a chosen color
- `fill`: Fills the arrow tip, which can be different from color, and can also be `fill=none`, so the tip looks open. `open` is an alias for `fill=none`

By putting arrow options into edge options, we get nested square brackets. That causes a headache for a compiler, so we have to protect the inner square bracket expression with curly braces. The easy `->` syntax becomes an expression like `-{tipname[options]}`. This is best seen with an example.

We take the code from *Figure 4.1* and add a dazzling edge with a flashy arrow. The edge will be very thick and colored in red, and the arrow will be orange, filled red, and big; so, we also add width and length:

```
\draw (tex) edge[very thick, draw=red,
  -{Stealth[color=orange, fill=red,
            width=8pt, length=10pt]} ]
  (pdf);
```

Figure 4.11 – A customized arrow tip

Agreed, that looks a bit crazy, but it is for instructional purposes, and you get the point: you can customize arrows in many ways apart from choosing predefined arrow tips.

In the following chapters, we will use arrows very often, so we will learn even more details about them.

Using the to operation

Remember, TikZ calls the instructions on a path an **operation**. Examples were the `node` and `edge` operations. What we draw with `edge` is not part of the main path. So, each edge can have its own appearance regarding styles or arrows, which makes it very flexible.

While the main focus of the current chapter is on the `edge` operation, there's another similar and handy one: the `to` operation. You will see it in examples on the internet and in the documentation, so let's also discuss it here, briefly at least.

The `to` operation can also be used to draw lines, curves, and arrows between nodes. `to` works with the current path options, such as color and arrow style. On the other hand, `edge` inherits the main path options but can take more path options in addition, such as its own color and arrow style. Suppose you don't need that many different path options. In that case, you can stick with `to`, which still understands the connection-specific options from the previous sections, such as `bend`, `in`, `out`, `looseness`, `relative`, `distance`, and their variations. Otherwise, you can use it in the same way.

For example, take this code line:

```
\draw[->] (tex) to (pdf);
```

It draws an arrow from the `tex` node to the `pdf` node, exactly as in *Figure 4.1*.

The `to` operation understands the connection-specific options, such as here:

```
\draw[->] (tex) to[out=45,in=225,looseness=1.5] (pdf);
```

That gives a curvy line with an arrow, going out from the `tex` node at an angle of 45 degrees, making a curve, so it's going to the `pdf` node at an incoming angle of 225 degrees:

Figure 4.12 – A curvy line with an arrow tip

But writing `to[->]` doesn't produce an arrow because `[->]` is a path option, like `color` and `thick`, for example.

The `edge` operation provides both path options and connection-specific options, so it's the most flexible approach.

The `edge` equivalent for *Figure 4.12* is as follows:

```
\draw[->] (tex) edge[out=45,in=225,looseness=1.5] (pdf);
```

In contrast to the `to` operation, you can move the arrow path option also to the `edge`:

```
\draw (tex) edge[out=45,in=225,looseness=1.5,->] (pdf);
```

The second code line is the better choice here. When creating edges, we should specify the arrow tip as an edge option instead of the \draw command because an edge is a separate path. Try this command:

```
\draw[->] (0,0) edge (1,0);
```

The following happens when you compile it:

- You get an arrow tip at the edge from (0,0) to (1,0) since the edge inherits the \draw options.
- You get an arrow tip right at the point of (0,0) because that's the main path.
- Look at the output to believe it:

Figure 4.13 – An undesired arrow tip

Just remember this when you create edges and you see an unexpected arrow tip.

Summary

In this chapter, you learned how to use edges to connect nodes and how to add text to them. You also extended your knowledge about using styles.

With what you now know about TikZ nodes, edges, and arrows, you can create complex diagrams with text elements, coloring, and further customizations.

In the next chapter, you will get a deeper understanding of TikZ styles to create more impressive drawings with less effort.

Further reading

The topics of this chapter are covered in depth in the following sections of the TikZ manual at https://texdoc.org/pkg/tikz:

- *Part III, Section 16, Arrows,* has a complete reference of all options and kinds of arrow tips with tabular overviews of the look of arrow tips and customizations. The online manual link is found here: https://tikz.dev/tikz-arrows.
- *Part III, Section 17.12, Connecting Nodes: Using the Edge Operation,* describes the edge and quotes syntax. It's part of the online manual section at https://tikz.dev/tikz-shapes.
- *Part V, Section 74, To Path Library,* explains the options for the to operation that can also be used with edge. You can read it online at https://tikz.dev/library-edges.

There are online resources worth taking a look at:

- TikZ drawing examples with arrow features: https://tikz.net/tag/arrows.
- TikZ examples working with arrows and creating special arrows: https://texample.net/tikz/examples/feature/arrows.
- An excellent tutorial about using and customizing arrows: https://latexdraw.com/exploring-tikz-arrows. The online examples are great starting points for your TikZ drawings.

5
Using Styles and Pics

In the previous two chapters, we learned about styles and used them with nodes, edges, labels, and pins. Now, we will take a closer look at styles and how to use them efficiently. Furthermore, we will deal with mini TikZ pictures that can be used as building blocks within a drawing.

Our main topics are the following:

- Understanding styles
- Defining and using styles
- Inheriting styles
- Using styles globally and locally
- Giving arguments to styles
- Creating and using pics

By the end of this chapter, you will know how to draw pictures more efficiently without repetitive syntax.

Technical requirements

As with every chapter, you need to have a LaTeX installation, including TikZ, or you can work online with Overleaf or work with the code at `https://tikz.org/chapter-05`.

The code is also available on GitHub at `https://github.com/PacktPublishing/LaTeX-graphics-with-TikZ/tree/main/05-using-styles-and-pics`.

In this chapter, we will use the `positioning` and `scope` TikZ libraries and the `tikzlings` package.

Understanding styles

We already customized nodes and edges using several `key=value` options within square brackets. Examples of **keys** are `color`, `shape`, `width`, and `font`.

We can define our own `key=value` sets. In TikZ, we commonly call such a set a **style**; and we give it a name. The name itself is also called a **key** in TikZ. The benefit for us is that such styles can contain a lot of various `key=value` settings and even code snippets.

Using styles is like working with macros in LaTeX. We can compare styles and macros in this way:

- If we have code that we use several times, we create a macro in LaTeX. If we have graphical properties values that we use several times, we create a named style in TikZ.
- Macros in LaTeX separate formatting from the content. Styles in TikZ separate graphical properties from the content of a drawing.
- Macros and styles save us from repeating code and help structure our documents and drawings.

Named keys have properties, such as a style, and codes. They can be set or modified using so-called **handlers**. The name of a key handler starts with a dot to distinguish them from regular key names.

That was quite theoretical, so we got to know some notations used in the TikZ manual. In the *Further reading* section, you will get references and links to thorough explanations of keys and handlers.

Let's now get practical and start drawing a picture with styles.

Defining and using styles

At first, we take the example of a node and its style. Let's take this node, which we call A:

```
\node (A) {A};
```

Well, it simply prints an A in the default font, without any shape or color. We change that now: let's have sans-serif and bold font, white text color, the shape of a circle, and color the circle like a blue ball:

```
\node [font = \sffamily\bfseries, text = white,
       shape = circle, ball color = blue] (A) {A};
```

That gives us a much fancier A:

Figure 5.1 – A fancy node

That's quite a lot of options for that node. If we have several nodes in a document, we don't want to repeat this for every single node. In *Chapter 3, Drawing and Positioning Nodes*, we saw the `every node/.style` syntax for applying such a set of options to all nodes in a drawing. That doesn't help us when we have different kinds of nodes in a drawing.

Let's explore this with the example of creating a **graph**. There's a famous problem in mathematics called *The Seven Bridges of Königsberg*. The city of Königsberg, now called Kaliningrad, had a river that separated the city into two parts, and the river had two islands. Seven bridges connected all of them. The challenge was to walk through the whole city while crossing each of those bridges exactly once.

Figure 5.2 visualizes the river with the two islands and the seven bridges. Try to imagine such a walk here:

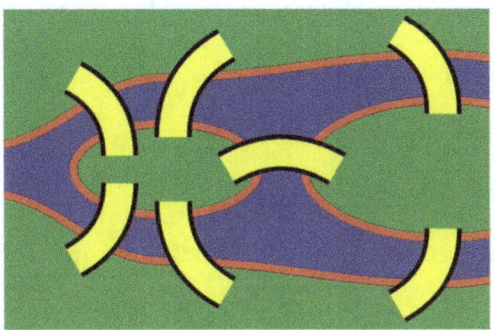

Figure 5.2 – The Seven Bridges of Königsberg

When the mathematician Leonhard Euler proved that this walk was impossible, he introduced the notation of a graph. We now call the nodes in this graph **vertices** to avoid mixing them up with nodes in general. Each vertex represents a piece of land. Each edge of the graph represents a bridge.

Now, we want to draw a graph with four vertices representing the two mainland parts and the two islands. We will add seven edges representing the bridges. Finally, we will number the edges from 1 to 7. You may already look ahead to *Figure 5.4* to see the target we want to achieve.

We will create styles to avoid repeating so many options, as we did in *Figure 5.1*. First, we create a vertex style that looks like the node in *Figure 5.1*. We will use the `\tikzset` command for this, which sets a TikZ `style` handler. This can be used both inside and outside the `tikzpicture` environment, even in the document preamble:

```
\tikzset{vertex/.style = {font = \sffamily\bfseries,
    text = white, shape = circle, ball color = blue}}
```

We'll make our lives easy and won't worry about the keys and handlers naming conventions; we call this highlighted syntax `vertex/.style` a **style** and modify it using the `\tikzset` command.

In the same fashion, we create a `bridge` style. We make it thick, color it yellow, and add a double-edge style:

```
\tikzset{bridge/.style = {thick, double = yellow,
    double distance = 1pt}}
```

Finally, we create a `number` style. The numbers will be nodes filled in red and with the same font as in the vertices:

```
\tikzset{number/.style = {font = \sffamily\bfseries,
  text = white, draw, fill = red}}
```

The node in *Figure 5.1* can now be reproduced simply by writing the following:

```
\node[vertex] (A) {A};
```

We place a second vertex node, B, to the right of A:

```
\node[vertex, right = 4 cm of A] (B) {B};
```

Now we can draw an edge between A and B with the bridge style and with a node displayed in the number style:

```
\draw (A) edge[bridge] node[number] {1} (B);
```

This gives us the start of our graph, with our first two vertices connected by a numbered bridge edge in the middle:

Figure 5.3 – A small graph

There's a variation of the /.style syntax: when we write /.append style, we can append a style to an existing style. For example, we can modify our bridge style to bend it in a direction:

```
\tikzset{bridge/.append style = {bend right}}
```

We can even do this in the middle of a `tikzpicture` environment.

We can define several styles with a single `\tikzset` command, separated by commas.

Putting everything we've learned together, we can now draw a complete *Königsberg bridges* graph. Here's the complete source code to see everything in one place:

```
\documentclass[border=10pt]{standalone}
\usepackage{tikz}
\usetikzlibrary{positioning}
\tikzset{
  vertex/.style = {font=\sffamily\bfseries, text=white,
         shape = circle, ball color = blue},
```

```
    bridge/.style = {thick, double = yellow,
            double distance = 1pt},
    number/.style = {font=\sffamily\bfseries, text=white,
            draw, fill = red}}
\begin{document}
\begin{tikzpicture}
  \node[vertex] (A) {A};
  \node[vertex, right = 4 cm of A] (B) {B};
  \draw (A) edge [bridge] node [number] {1} (B);
  \node[vertex, below = 2cm of A] (C) {C};
  \node[vertex, above = 2cm of A] (D) {D};
  \tikzset{bridge/.append style = {bend right}}
  \draw (C) edge [bridge] node [number] {2} (B)
        (B) edge [bridge] node [number] {3} (D)
        (C) edge [bridge] node [number] {4} (A)
        (A) edge [bridge] node [number] {5} (C)
        (A) edge [bridge] node [number] {6} (D)
        (D) edge [bridge] node [number] {7} (A);
\end{tikzpicture}
\end{document}
```

When you compile this document, you will get the following comprehensive graph:

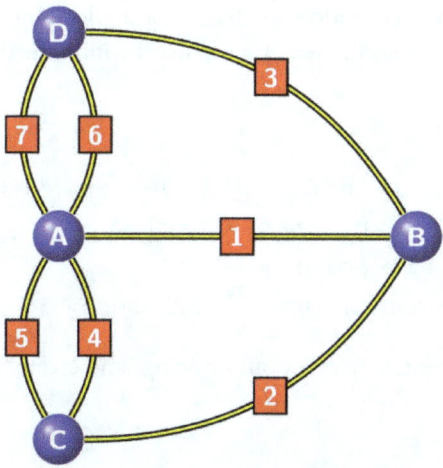

Figure 5.4 – A graph for the Seven Bridges of Königsberg problem

The /.append style handler is helpful when you want to add a key and value to a global style just locally in a picture, or when you want to add some value for a key to a predefined style without knowing that style's details.

What you append overwrites already existing keys. Let's say you have a style that contains `fill=blue`, and you append `fill=red`; then, it will be filled red.

There might be occasions when we want it to work the other way round. We can, for example, prepend `fill=red` so we get a red filling only if the original style has no `fill` key that would override red. The handler for this is `/.prefix style`, used like this:

```
\tikzset{vertex/.prefix style = {fill=red}}
```

Now that we can use styles, we have improvements in our code:

- We can separate formatting from drawing content
- We can reuse styles; we can even use the same styles in other drawings
- We get shorter code without much repetition
- We benefit from a clear and descriptive notation of drawing elements

To improve reuse and avoid repetition, we can define new styles based on previously defined styles. That's the topic of our next section.

Inheriting styles

You noticed that we used the commands for sans-serif and bold font and white text color both in the vertex and number styles. To avoid repetition and to have a single point of definition, we can define a style used by both. Let's call it `mytext`. Then, we can use it within the definitions of both `vertex` and `number`:

```
\tikzset{
  mytext/.style = {font=\sffamily\bfseries, text=white},
  vertex/.style = {mytext, shape=circle,
                   ball color = blue},
  number/.style = {mytext, draw, fill=red}}
```

That way, we can define fundamental styles for our drawings and create further specific styles based on them.

Similarly, we can define specific styles based on other styles, such as highlighting elements in a drawing. Here, we define a general `highlight` style and combine it with other styles:

```
\tikzset{highlight/.style = {draw=yellow, very thick,
        densely dotted},
      highlight vertex/.style = {vertex, highlight},
      highlight number/.style = {number, highlight}}
```

Then, we adjust the lines for the `A` node and the `1` edge accordingly:

```
\node[highlight vertex] (A) {A};
\draw (A) edge [bridge] node [highlight number] {1} (B);
```

Compile this, and the drawing will show the `A` node and the `1` edge highlighted:

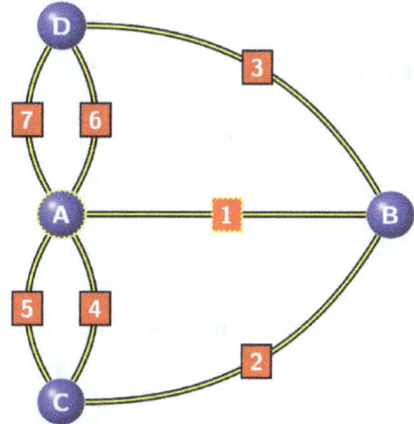

Figure 5.5 – Highlighting vertex A and edge 1

Now, let's briefly look at alternative ways to define styles, focusing on keeping it local to a single picture or environment.

Using styles globally and locally

Using the `\tikzset` command, you can define styles *globally* for your whole document. This is especially useful when you have several similar drawings in your document. For example, in a book about graph theory, you` probably want to have the same styles for vertices, edges, and labels in all drawings throughout the book, so it's good to use `\tikzset` in the preamble.

In older documents, you will see the `\tikzstyle` command with the following form:

```
\tikzstyle{my style} = [options]
```

That command is deprecated and should not be used anymore, according to the TikZ creator, so bear this in mind when you see it in older code on the internet.

In situations where styles between drawings are different, it can be preferable to define styles *locally*, so they are only valid in a single picture. That can be done by setting the styles as options in the `tikzpicture` environment. For example, if you have a drawing where you want to have a particular local `vertex` style, you might do the following:

```
\begin{tikzpicture}[vertex/.style = {shape = circle,
                                     ball color = blue}]
\node[vertex] (A) {A};
\end{tikzpicture}
```

In another picture, you can then define a different `vertex` style.

You can restrict the effect of styles and options even to just a part of the picture using a `scope` environment:

```
\begin{scope}[thick, draw=red]
...
\end{scope}
```

Here, everything that's within the `scope` environment will have thick lines and be drawn in a red color. Once the environment ends, it's neither thick nor red anymore. Scopes are used to apply settings to a whole piece of TikZ code. Again, it's for reducing repetitions and structuring a TikZ picture.

We will use scopes in *Chapter 7, Filling, Clipping, and Shading*, for restricting clipping settings, and in *Chapter 9, Using Layers, Overlays, and Transparency*, for delimiting what we write on different layers of a drawing.

For very small scopes, such as a single command, there's a shortcut called `\scoped`. Instead of the preceding `scope` environment, we could write the following:

```
\scoped[thick, draw=red]{\draw ...}
```

This makes sense especially when we have to use scoping to set an option not directly supported by a command, such as clipping and layers, as we'll see later in the aforementioned chapters.

Another shortcut is simply to use curly braces and square brackets. To be able to use this syntax, you first need to load the `scopes` library:

```
\usetikzlibrary{scopes}
```

Now, you can begin a scope by writing an opening curly brace followed by square brackets with options. With a closing curly brace, the scope ends. Our previous `scope` example is even shorter now:

```
{[thick, draw=red]
   ...
}
```

This short syntax of curly braces is parsed and detected only if the following occurs:

- The opening brace is immediately followed by options in square brackets
- You start it after a semicolon ends a path, after a previous scope has ended, or at the beginning of a picture or a scope

Otherwise, the braces will be treated like standard TeX braces.

Similar to environments and macros, we can create styles that have parameters, also called **arguments**. We will do this in the next section.

Giving arguments to styles

Remember that in *Figure 5.5*, we defined the `vertex` style in the following way:

```
\tikzset{vertex/.style = {mytext, shape = circle,
  ball color = blue}}
```

We can introduce an argument when we intend to have different colors with the same style. One argument is easily supported; we can write the following, similar to arguments in macros:

```
\tikzset{vertex/.style = {mytext, shape = circle,
  ball color = #1}}
```

Now, we can change our code for *Figure 5.5* to choose colors as arguments:

```
\node[vertex=blue] (A) {A};
\node[vertex=green, right = 4 cm of A] (B) {B};
```

So, `#1` represents an argument in our style, and with `style=value`, we set that value for `#1`. We can specify a value that's used when no value is given using the so-called `.default` handler:

```
\tikzset{vertex/.default=blue}
```

Now, we can write `\node[vertex]` for a blue node by default, and `\node[vertex=green]` for a green node.

We may write `style={value}` to avoid misunderstandings, especially when it comes to two or more arguments. Then we have to provide them as follows:

```
style = {value1}{value2}
```

This is better explained with another example. We will create a style with two arguments.

For our network diagram in *Figure 3.25*, we created nodes with images. We added the images using `\includegraphics` in the node text, like this:

```
\node (router) [inner sep=0pt]
  {\includegraphics[width=2cm]{router.pdf}};
```

We prefer to avoid repeating this when we have another router in our drawing. Since we now know more about styles, we will now define styles for nodes with images.

For this, we will use the `path picture` syntax, which allows us to add code as an option to a path. Consider this snippet:

```
path picture = <some code>
```

When we give this option to a path, then after the path has been drawn and filled, the code will be executed. The drawing that the code produces will be clipped to the path. In that code, we can have `\draw` commands and `\node` commands, for example.

Before we use it, we need to know the syntax for a style with two arguments. Basically, for a style called `image`, it is as follows:

```
image/.style 2 args = <some code with #1 and #2>
```

Agreed, this is still too dry. Let's combine it to create an image style with two arguments that include a node with an image:

```
\tikzset{
  image/.style 2 args = {path picture = {
    \node at (path picture bounding box.center) {
      \includegraphics[width=#1cm] {#2}};}}}
```

With that, we come close. This style adds an image of a certain width (the first argument), and with a particular filename (the second argument).

Now, we can add styles for various images as follows, within `\tikzset`, of course:

```
router/.style = { image = {2}{router.pdf} },
switch/.style = { image = {3}{switch.pdf} },
```

This gives us node styles for routers with 2 cm width and switches with 3 cm width that we can repeatedly use in our drawings like this:

```
\node[router] (r) {};
\node[switch, right = 4 cm of r] (s) {};
```

So, we have a router node, (r), and a switch node, (s). Using such styles, we can add many nodes with the same images to our drawing and have a node style for each image.

We can use even more arguments. This n args syntax allows us to use multiple arguments, for example, five arguments for a style name:

```
name/.style n args={5}{some code}
```

We can have from zero to nine arguments. It's good to know that we have the possibility of styles with many arguments, but actually needing to use that many is rather rare. The .default handler can also be used with several arguments; list them separated with curly braces, like this:

```
image/.default = {2}{example.pdf}
```

By using styles like sets of element properties, we can use whole partial drawings as building blocks, which will be our next topic.

Creating and using pics

In LaTeX, we can write macros containing code that can be used repeatedly. How about using TikZ picture code repeatedly in a drawing? We cannot simply put one tikzpicture environment into another one. These pictures and their elements would interfere with each other's styles and settings.

To solve this, TikZ provides a syntax for creating small pictures that can be used as building blocks in a TikZ drawing. The feature name is pic; let's also call these short pictures *pics*.

A pic is a TikZ drawing code sequence, defined in a similar way to setting a style. To get practical, we will define a smiley pic based on the code for our self-made smiley in *Chapter 2, Creating the First TikZ Images*. The basic syntax is as follows:

```
\tikzset{smiley/.pic={ ... drawing commands ... }}
```

Like .style, .pic is also an example of a **key handler**.

We take our code for *Figure 2.11* and put this code into the \tikzset command in the following way:

```
\tikzset{smiley/.pic={
  \draw[shading=ball, ball color=yellow] (0,0)
    circle [radius=2];
```

```
\draw[shading=ball, ball color=black] (-0.5,0.5,0)
    ellipse [x radius=0.2, y radius=0.4];
\draw[shading=ball, ball color=black] (0.5,0.5,0)
    ellipse [x radius=0.2, y radius=0.4];
\draw[very thick] (-1,-1) arc [start angle=185,
    end angle=355, x radius=1, y radius=0.5];}}
```

To use the smiley pic in our TikZ pictures, we can place it like a node. We can write together with \draw:

```
\draw pic {smiley};
```

We can add the usual drawing options such as scale and rotate and give coordinates for the position. Without coordinates, (0,0) is implicitly used.

Consider this code to understand the positioning:

```
\draw (2,4) pic {smiley};
```

Here, the smiley pic is positioned in the picture so that the origin (0,0) within the pic is placed at the coordinates (2,4) in the surrounding TikZ picture. In contrast to a node, a pic doesn't have anchors.

Now, we can add smileys in all our TikZ pictures quickly and with ease, saving us work. Or we can be a bit funny and put many smileys in a single picture. Let's do this just for practice. We put smiley pics at various coordinates with different size scaling and rotation:

```
\begin{tikzpicture}
\draw      pic {smiley}
   (2,2)    pic [scale=0.5,  rotate=-30] {smiley}
   (-2,1.5) pic [scale=0.3,  rotate= 30] {smiley}
   (-1.6,2) pic [scale=0.15, rotate=-20] {smiley}
   (0,2)    pic [scale=0.2,  rotate=-10] {smiley};
\end{tikzpicture}
```

When you compile this picture in a document, you will get the following:

Figure 5.6 – Repeating small pictures

If you repeatedly use specific drawing codes, you can store them as pics in a separate .tex file and input them within your preamble, or you can create a .sty file for this. There are even TikZ packages that provide ready-to-use pics.

We will try the tikzlings package. That package contains drawings of many cute animals, called **TikZlings**, which we can add and position as pics in our drawings.

First, we load the package:

```
\usepackage{tikzlings}
```

Now we have a library of more than 20 animals to hand that we can use with \pic{animal} or \draw pic{animal} and position them exactly where we want. Let's try that, and to work with positioning, we will also use the grid from *Chapter 2*, *Figure 2.1*. Since we have learned how to use it now, we will create a pic for the grid. Let's call it mygrid:

```
\tikzset{mygrid/.pic = {
    \draw[thin, dotted] (-3,-3) grid (3,3);
  \draw[->] (-3,0) -- (3,0);
  \draw[->] (0,-3) -- (0,3);}}
```

Now we are ready for the picture. We chose the following animals: chicken, pig, bear, penguin, and owl:

```
\begin{tikzpicture}
\draw       pic {mygrid}
   (-1,0)   pic {chicken}
   (1,0)    pic {pig}
   (-2,-2)  pic {bear}
   (0,-2)   pic {penguin}
   (2,-2)   pic {owl}
```

```
;
\end{tikzpicture}
```

We used all pics in a single \draw command. When we compile it, we get this picture:

Figure 5.7 – Mesmerizing animals on a grid

Since we know the grid dimensions, we see that the pics' origins are a bit below the middle of each animal's feet.

The TikZlings are very famous in the LaTeX and TikZ communities. Like various TikZ-related packages, it started just as a bit of fun. But more seriously, programming such things for fun is also educational. Just look at the source code of TikZlings, for example, at https://github.com/samcarter/tikzlings. It's a challenging piece of TikZ code that allows customizing those cute animals via style options in many different ways, such as color and 3D effects, and you can also add accessories including hats and clothes, speech bubbles, shovels, brooms, lightsabers, pizza, and drinks.

Summary

By working through this chapter, you have gained expertise and a professional workflow with which to create TikZ pictures systematically. You can now define and apply your own styles to drawing elements.

In the next chapter, we will apply our knowledge to trees and graphs.

Further reading

The TikZ manual has the complete reference on keys, handlers, styles, and pics. Open it using texdoc tikz at the command line or visit https://texdoc.org/pkg/tikz.

The topics of this chapter are covered in depth in these sections in *Part III*:

- *Section 12, Hierarchical Structures,* talks more about scopes. The direct online link is `https://tikz.dev/tikz-scopes`.
- *Section 18, Pics: Small Pictures on Paths,* describes the `pic` feature. The online manual link is `https://tikz.dev/tikz-pics`.

Part VII, Section 87, Key management, explains keys and handlers. An online version is at `https://tikz.dev/pgfkeys`. It also describes the usage of style arguments in depth.

While this chapter gives you a quick start for easy comprehension, the TikZ manual is the complete reference.

There are further online resources worth taking a look at:

- `https://tikz.net/tag/styles` shows TikZ drawing examples focusing on styles
- `https://texample.net/tikz/examples/feature/styles` also has TikZ examples with a focus on styles

The TikZlings also have their own manual with more than 60 pages. You can open it using `texdoc tikz` or visit `https://texdoc.org/pkg/tikzlings`.

6
Drawing Trees and Graphs

With nodes, edges, and styles, we can already build complex pictures. To make our work easier, TikZ provides libraries with support for various common types of graphics.

Each library provides specific drawing commands and corresponding style options.

In this chapter, you will learn to work with TikZ libraries for the following topics:

- Drawing trees
- Creating mind maps
- Producing graphs
- Positioning in a matrix

By the end of the chapter, you will be able to efficiently produce such types of graphics.

Technical requirements

Apart from your local LaTeX installation, you can work with the code at `https://tikz.org/chapter-06` or use Overleaf for online compiling.

The code examples are available on GitHub at `https://github.com/PacktPublishing/LaTeX-graphics-with-TikZ/tree/main/06-drawing-trees-graphs-charts`.

In this chapter, we will use the `trees`, `graphs`, `matrix`, and `quotes` libraries. Furthermore, we will use the `hvlogos` package for printing LaTeX-related logos.

Drawing trees

Trees are a specific kind of graph where any two vertices are connected by exactly one path, which consists of edges. We may encounter them in mathematics, such as in graph theory, in computer science, or in any drawing that illustrates parent-children relationships in a hierarchical way.

We already know the `node` and `edge` operations, and the `child` operation combines both of them. Specifically, it connects a parent node and one or more child nodes by edges, each from the parent to a child. Moreover, TikZ counts the children and determines an appropriate positioning for each of them.

A very simple example is this:

```
\node {A} child { node {1} edge from parent };
```

The output is this very simple tree:

Figure 6.1 – A simple tree

`edge from parent` is a special path operation that adds an edge from the parent to the child. We can add options and nodes to this edge, as we know from *Chapter 4, Drawing Edges and Arrows*. For example, let's make this edge a dashed arrow with a short piece of text along it:

```
\node {A} child { node {1}
   edge from parent [dashed, ->]
   node[above, sloped, font=\tiny] {down} };
```

That changes our diagram in this way:

Figure 6.2 – A customized edge in a tree

If we need just a simple edge, we can edge from the parent, and TikZ adds it implicitly to the child. So we can write the code for *Figure 6.1* much shorter this way:

```
\node {A} child { node {1} };
```

That's not impressive yet. The next useful feature of TikZ trees is that we can have several children. We simply state them, then TikZ counts them, calculates their placement, and draws them with edges.

Let's have five children right away:

```
\node {A}
  child { node {1} }
  child { node {2} }
  child { node {3} }
  child { node {4} }
  child { node {5} }
;
```

That code results in this well-balanced tree:

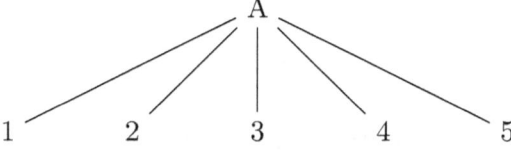

Figure 6.3 – A tree with five children

Since these are nodes and edges, we can add style options to them like we used to do, such as in square brackets for each node. To avoid repetition, we can set style options for child paths and for the whole tree. It works this way:

- **For the whole tree**: Specify the options for the tikzpicture environment
- **For the parent of all, which we call the root node**: Give options to that node, as you already know, in square brackets
- **For all children**: Give them in square brackets between the root node and the first child node
- **For a certain child path**: Add them in square brackets to the child operation
- **For a specific node in the tree**: Add as an option to the node within the child path

To better see it, look at this code where we play with the options:

```
\begin{tikzpicture}[thick]
  \node [draw, black, rectangle] {A}
    [red, ->]
    child { node {1} }
    child { node {2} }
    child [densely dashed]
      { node [draw, blue, circle] {3} }
    child { node {4} }
    child { node {5} }
  ;
\end{tikzpicture}
```

The highlighted options change the diagram from *Figure 6.3* to this one:

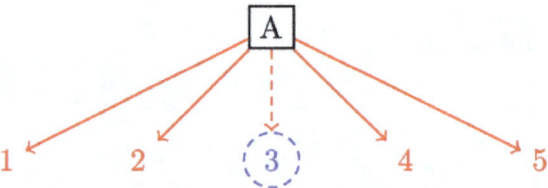

Figure 6.4 – A tree with custom style options

You can also see that the style options are inherited. The dashed child 3 still has a red arrow like all the other children, and node 3 is thick like the whole tree. In addition, node 3 has its own style options, being blue with a circular border, still inheriting the `densely dashed` option from its child path.

There are further layout options for distances between parents and children and between neighbor children, which is especially useful when we have several levels in a tree. Child nodes can also act as parents and can have further children. Let's try this with another example.

For practicing tree building and exploring further features, we will build a tree of TeX and LaTeX relationships, including ConTeXt and several LaTeX engines. For writing TeX-related logos, we will load the `hvlogos` package.

This shall be our starting point, with the tree code and the additional package highlighted:

```
\documentclass[border=10pt]{standalone}
\usepackage{tikz}
\usepackage{hvlogos}
```

```
\begin{document}
\begin{tikzpicture}
  \node {\TeX}
    child { node {\LaTeX} }
    child { node {\ConTeXt} }
  ;
\end{tikzpicture}
\end{document}
```

Compile this minimal example and you get this output:

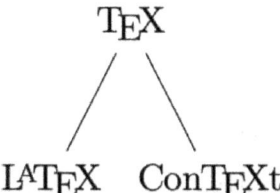

Figure 6.5 – A minimal TeX relationship tree

Now let's add a third level. The LaTeX node shall have three children, representing the LaTeX format based on the engines pdfTeX, XeTeX, and LuaTeX. The new nodes are children of the LaTeX node. So we have to use the child operation within the LaTeX child path, which means we have to nest the syntax. Here, the addition is highlighted:

```
\node {\TeX}
  child { node {\LaTeX}
    child { node {\pdfLaTeX} }
    child { node {\XeLaTeX} }
    child { node {\LuaLaTeX} }
  }
  child { node {\ConTeXt} }
;
```

When we compile the code, we get this now:

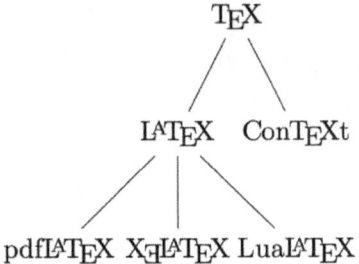

Figure 6.6 – A TeX and LaTeX relationship tree

The first level after the root looks okay, the second level looks cramped, and the vertical distance is high compared to the children's spacing.

We have two kinds of distances that we can adjust:

- `level distance` for the vertical distance between parent and child
- `sibling distance` for the horizontal distance between neighbor children

Similar to styles in trees, as we saw previously, we can set values for the whole tree as options for the `tikzpicture` environment, valid for all children between the root node and the first child, or as options for specific children. Furthermore, we can apply it as a `level x` style to `tikzpicture`, like this:

```
\begin{tikzpicture}[
  level 1/.style = { level distance  = 8mm,
                     sibling distance = 20mm },
  level 2/.style = { level distance  = 10mm,
                     sibling distance = 20mm } ]
...
\end{tikzpicture}
```

Adding this to our code example, we get this adjusted tree:

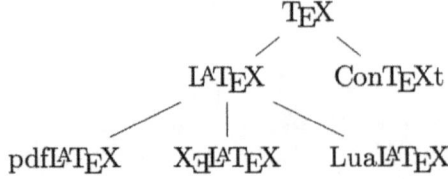

Figure 6.7 – Adjusted distances in a tree

Let's say we want to give the nodes a nice appearance and add labels to the edges. We could use the way that we already know, which is setting `every node/.style` to the desired values; however, this also affects the label nodes. We can use `every child node/.style` for this.

In the following code, we define a general tree node style that we apply to the root node and every child, and a style for the labels we call `engine`. The `engine` labels shall stand for the compiling engine for the LaTeX version:

```
\begin{tikzpicture}[
  level 1/.style = { level distance = 8mm,
                     sibling distance = 20mm },
  level 2/.style = { level distance = 10mm,
                     sibling distance = 20mm },
  treenode/.style = {shape = rectangle,
    rounded corners, draw,
    top color=white, bottom color=blue!30},
  every child node/.style = {treenode},
  engine/.style = {inner sep = 1pt, font=\tiny, above}
]
\node [treenode] {\TeX}
  child { node {\LaTeX}
    child { node {\pdfLaTeX}
      edge from parent node[engine, sloped] {\pdfTeX}}
    child { node {\XeLaTeX}
      edge from parent node[engine, left]   {\XeTeX} }
    child { node {\LuaLaTeX}
      edge from parent node[engine, sloped] {\LuaTeX}}
  }
  child { node {\ConTeXt} }
;
\end{tikzpicture}
```

When we compile this example, we get the following:

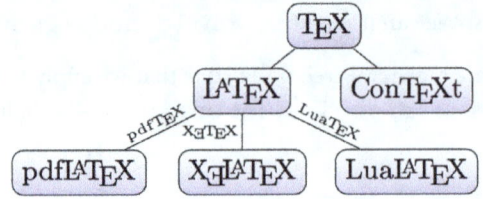

Figure 6.8 – A tree with styles and labels on edges

While in computer science, trees are often top-down, we can choose a different layout. By adding the grow=right option to the tikzpicture environment, we get this tree, which goes left to right:

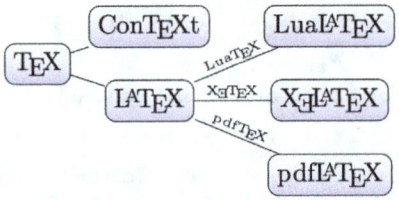

Figure 6.9 – A tree from left to right

Of course, we choose the different levels and sibling distances when we turn the tree. For grow, you can choose among down, up, left, right, north, south, east, west, north east, north west, south east, and south west. Furthermore, you can set an angle in degrees, such as grow=90, which is the same as grow=up.

If you would like to reverse the order of how TikZ draws the children, use grow' instead. For example, when we write \begin{tikzpicture}[grow=up], *Figure 6.3* changes to this:

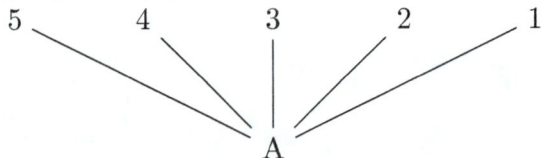

Figure 6.10 – A tree growing up

The order is still clockwise, like in *Figure 6.3*, but it's not left to right anymore. That may be not very clear. When we change the option to grow', we get this:

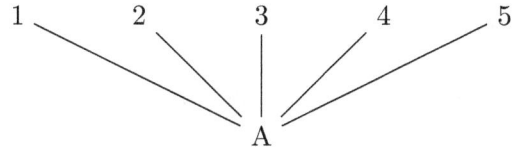

Figure 6.11 – A tree growing up with children in inverted order

The order is now counterclockwise, but we see it again as left-to-right.

The trees library offers additional features. You can load it as usual:

```
\usetikzlibrary{trees}
```

The most useful is the cyclic child nodes arrangement:

- grow cyclic switches to the positioning of children around the parent node with a fixed distance, like on a circle. The circle's radius is level distance, which we already used before.
- sibling angle = ... defines the angle between two siblings in degrees.
- clockwise from = ... is a value in degrees, where the first child will be placed, in the level distance. The second child is placed with sibling angle from the first child and the same level distance, and so on. In this case, you don't need to add grow cyclic.
- counterclockwise from = ... is like the previous option in the other direction.

You can apply these options to the tikzpicture environment. It's good to see an example, so take the code from *Figure 6.3* and use this:

```
\begin{tikzpicture}[clockwise from = 180,
  sibling angle=45]
```

The diagram changes to this arrangement:

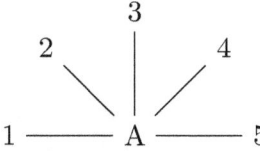

Figure 6.12 – A tree with a circular child node arrangement

In the next section, we will use this feature when we draw mind maps, which are trees where topics and subtopics are displayed in hierarchical circles.

Creating mind maps

We already know about trees for organizing information hierarchically, usually growing top-down or left-to-right, with several levels. Imagine a tree with a root in the center, and the first level of children is circular around it. Each child again is surrounded by children of the next level.

Such a diagram is called a **mind map** and is famous for visualizing ideas. We have a central **concept**, from which child concepts branch out in various directions. Each child concept can again have children.

TikZ provides the `mindmap` library, which can display a root concept as a circle in the center and child concepts as smaller circles around it, connected by **branches**, which are edges.

Load it with `\usetikzlibrary{mindmap}`. Then, add the `mindmap` option to the `tikzpicture` environment and build a tree with children, like in the previous section. Just the nodes will have the `concept` style. That includes a `concept color` value that you can set.

But let's take small steps and do this with a very small example. We will create a mind map with a root concept node with huge bold text and a child concept node with a bit smaller text, white text in blue-filled nodes:

```
\begin{tikzpicture}[
    mindmap,
    concept color = blue!50,
    text = white,
  ]
  \node [concept, font=\Huge\sffamily\bfseries] {TikZ}
    child [clockwise from = 0] {
      node [concept, font=\Large\sffamily] {Graphs}
    };
\end{tikzpicture}
```

This code gives us the following:

Figure 6.13 – A minimal mind map

When we want every node to have the concept style, we can simply write nodes = {concept} in the picture options.

We can adjust the styles for each concept level, for example, by changing the font size. We should append such styles not to override the concept style itself. To practice this, our mind map should be a bit bigger.

We apply what we already know, nest child nodes as we know from trees, and add a font style. So, this is our code now:

```
\begin{tikzpicture}[
  mindmap,
  text = white,
  concept color = blue!50,
  nodes = {concept},
  root/.append style = {
    font = \Huge\sffamily\bfseries},
  level 1 concept/.append style =
    {font = \Large\sffamily, sibling angle=90},
  level 2 concept/.append style =
    {font = \normalsize\sffamily}
]
node [root] {TikZ} [clockwise from=0]
  child [concept color=blue] {
    node {Graphs} [clockwise from=90]
      child { node {Trees} }
      child { node {Mind maps} }
      child { node {DOT syntax} }
      child { node {Algorithms} }
```

```
    };
\end{tikzpicture}
```

Already a bit challenging! We have to ensure we close all the curly braces correctly and don't miss one. That code gives us this:

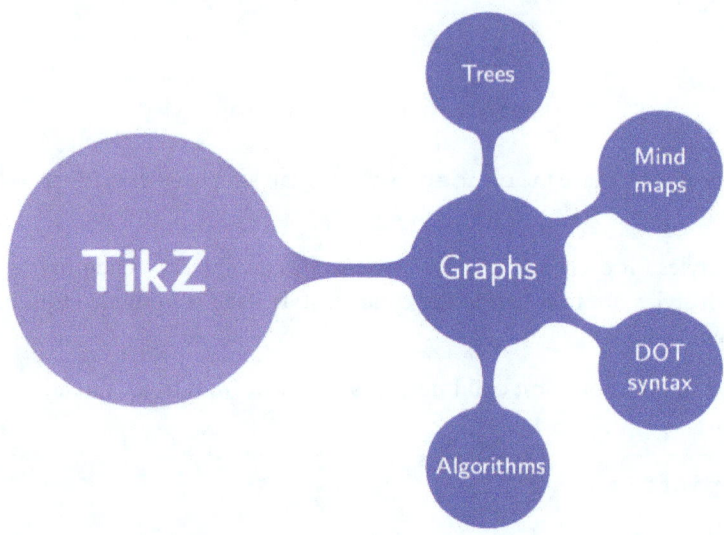

Figure 6.14 – A mind map with a root and two levels

This might seem familiar to you – it's a mind map of this chapter. You may also notice the nice transition of one concept color to the other concept color, implicitly done by the `mindmap` styles.

The default sibling angle is 60 degrees. We set it to 90 degrees just for the first level to have our first-level children at the right, bottom, left, and top of the root node when we now continue to add children. We will declare more children in the same way to visualize this book's content. Our adjustment for each child set is just `concept color` and the start angle:

```
child [concept color=green] {
  node {Basics} [clockwise from=30]
    child { node {Drawing} }
    child { node {Colors} }
    child { node {Nodes} }
    child { node {Edges} }
    child { node {Styles} }
}
```

The full code is on GitHub and on `TikZ.org`, as mentioned. We get this final mind map:

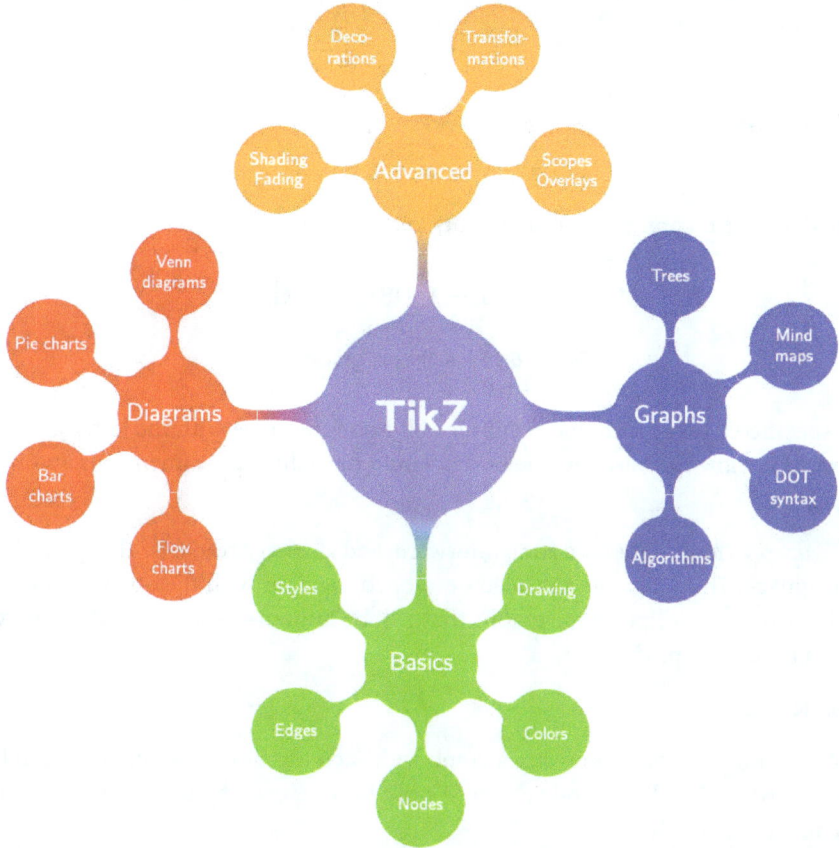

Figure 6.15 – A comprehensive mind map

There are more kinds of graphs than just trees and mind maps, so we will look at more general graphs in the next section. Specifically, we will learn a new drawing syntax.

Producing graphs

The syntax with child nodes and edges can feel lengthy, and having many curly braces may lead to small mistakes. TikZ provides a special syntax for graphs that is very concise.

To be able to use it, we have to load the `graphs` library with this command:

```
\usetikzlibrary{graphs}
```

This gives us a new command called \graph, which generates even complex graphs with short specifications. Here's a quick example of how it looks, representing a classic LaTeX compiling process:

```
\begin{tikzpicture}[ nodes = {text depth = 1ex,
    text height = 2ex}]
 \graph { tex -> dvi -> ps -> pdf };
\end{tikzpicture}
```

That highlighted \graph command produces this image:

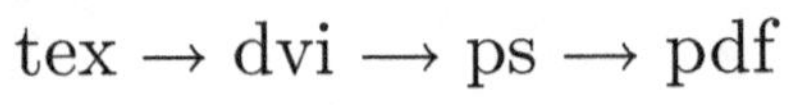

$$\text{tex} \to \text{dvi} \to \text{ps} \to \text{pdf}$$

Figure 6.16 – A simple graph

Note that we specified a text depth and height for all nodes because with the letter p in the node text, which extends below the baseline, the nodes would have different dimensions and would not be properly aligned.

The code for *Figure 6.16* looks short and straightforward, and also very readable. The syntax is inspired by the DOT language. This language has been developed together with the open source **Graphviz** package and is a simple description for graphs that will be human-readable. In other words, TikZ welcomes and supports Graphviz users.

Let's look at basic rules.

A sequence of text and -> shortcuts creates a graph with such text as nodes with arrows in between, like in *Figure 6.16*, which is called a **node chain**. We can have several chains in a graph, separated by commas or semicolons, like this:

```
\graph { tex -> dvi -> ps -> pdf,
         bib -> bbl,
         bbl -> dvi };
```

That produces this graph:

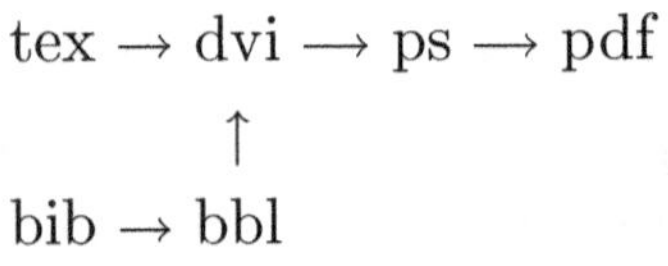

$$\text{tex} \to \text{dvi} \to \text{ps} \to \text{pdf}$$
$$\uparrow$$
$$\text{bib} \to \text{bbl}$$

Figure 6.17 – Node chains in a graph

You see two chains – one in each row. The third chain refers to already existing nodes, so it just adds an arrow between them.

By enclosing nodes or whole chains in curly braces, you get a **node group** or a **chain group**. Each node of such a group is connected with the previous node and the following node. Instead of nodes, you can have again small graphs in a group. As in this example:

```
\graph { tex -> {dvi, pdf } -> html };
```

That gives us this figure with `dvi` and `pdf` as a group, connected to the nodes before and after them:

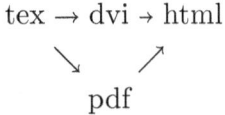

Figure 6.18 – Node groups

The following types of edges are available:

- `->` is an arrow to the right
- `<-` is an arrow to the left
- `<->` is a bidirectional arrow
- `--` is an edge without an arrow tip
- `-!-` means no edge shall be here

Every such edge can be followed by options in square brackets, as we know from edges.

All of this enables us to already create complex graphs.

Positioning can be the next challenge. Specifically, we may want to change the distance between nodes and levels. There are several keys for the `\graph` command:

- `grow up`, `grow down`, `grow left`, and `grow right`: Set the value as the distance between the centers of neighbor nodes on a chain in the growth direction
- `branch up`, `branch down`, `branch left`, and `branch right`: Set the value as the distance between siblings or adjacent branches

Here, we apply a distance of 2 cm between the node centers of the chain in *Figure 6.16*:

```
\graph [grow right = 2cm] { tex -> dvi -> ps -> pdf };
```

We get a much wider graph:

$$\text{tex} \longrightarrow \text{dvi} \longrightarrow \text{ps} \longrightarrow \text{pdf}$$

Figure 6.19 – Node distance in a graph

This allows us to add labels to the edges. Remember, we can use the `quotes` library:

```
\usetikzlibrary{quotes}
```

Now we can add labels as options in square brackets to the edges:

```
\graph [grow right = 2cm]
  { tex -> ["latex"] dvi
       -> ["dvips"] ps -> ["ps2pdf"] pdf };
```

The `every edge quotes` style is used here, so we can modify it, as an option to the `tikzpicture` environment, to have it in a very small typewriter font:

```
every edge quotes/.style = {font=\tiny\ttfamily,
   above, inner sep = 0pt}]
```

Our graph has edges with labels now:

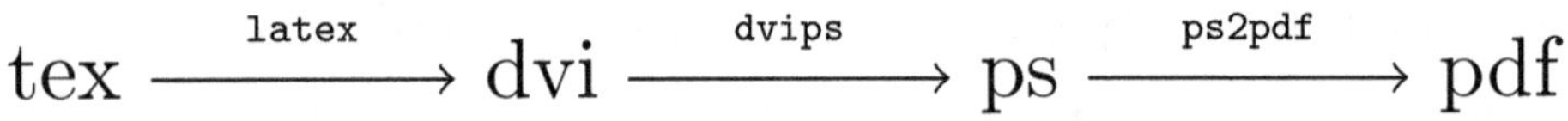

$$\text{tex} \xrightarrow{\text{latex}} \text{dvi} \xrightarrow{\text{dvips}} \text{ps} \xrightarrow{\text{ps2pdf}} \text{pdf}$$

Figure 6.20 – A graph with labels on edges

And, of course, apart from labels, we can add formatting options such as color or thickness to the edges in square brackets, like with the labels.

This exercise may not show the full power of the `graphs` syntax. We have a real benefit if we use a lot of graphs or huge graphs with dozens of nodes and edges, like in computer science or mathematics.

While TikZ arranges nodes in trees and graphs nicely and automatically, we may want to have more freedom in placement. The following section shows an efficient way.

Positioning in a matrix

Diagrams and drawings, in general, often have a rectangular structure, with elements or text arranged vertically and horizontally, like on a grid. TikZ offers a `matrix` node style for such a placement. Here is a very simple example:

```
\node [matrix, draw] {
   \node{A}; & \node{B}; & \node{C}; \\
   \node{D}; & \node{E}; & \node{F}; \\
};
```

This gives us the following rectangular node, with nodes placed in a matrix grid:

Figure 6.21 – A simple matrix node

The syntax is similar to LaTeX's `array` and `tabular` environments: columns are separated by &, and rows end with \\. Also note that the last row must end with \\.

Each cell can contain a node or a small drawing, or it can be left empty. TikZ adjusts the size of the cells automatically, so it fits the content.

Since the matrix is a node, you can apply what you already know and add shape and style options to the node in square brackets or give it a name in parentheses.

There's a short `\matrix` command, which is equivalent to `\node [matrix]` and saves some typing. Even more convenient, TikZ has a `matrix` library that provides the `matrix of nodes` option, where each cell is implicitly a node. So, we just need to insert the text in cells.

At first, we load the `matrix` library in our preamble:

```
\usetikzlibrary{matrix}
```

With the `matrix of nodes` option and the `\matrix` command, our code for *Figure 6.21* becomes much shorter and easier to write:

```
\matrix[matrix of nodes, draw] {
    A & B & C \\
    D & E & F \\
};
```

In our first matrix example, it was clear how to apply shapes and other options to nodes, as usual, in square brackets to the `\node` command. How can we do it here?

The first and easiest way is when all cell nodes within the matrix have the same style. In this case, we can set the `nodes` option as we did in the mind map earlier in this chapter.

The second way, for specific cells, is inserting options with vertical bars right before the cell content. All that's between the bars is passed to the implicit `\node` command in this cell.

For example, here, we define that all nodes in our matrix example shall have a `circle` shape and a `minimum width` of 2 em, and we color the last cell node in red:

```
\matrix [matrix of nodes, draw,
    nodes = {circle, draw, minimum width=2em} ] {
    A & B & C \\
    D & E & |[red]|F \\
};
```

Our diagram changes to this:

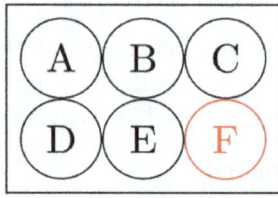

Figure 6.22 – Matrix cell nodes with style options

In *Figure 6.22*, we see that the cells are tightly close to each other. We can set values for row and column spacing this way:

- `row sep` is the value of the space inserted between rows
- `column sep` is the value of the space between columns

We will use it in our next example.

We can use the matrix features for building diagrams with such a convenient positioning syntax. Let's say we want to modify *Figure 6.16* to create a diagram that visualizes TeX input and output formats and explains that we can convert from all formats into PDF. We shall have a top row and the PDF format in the second row, with arrows between the formats depicting the conversion.

At first, we build the matrix. Note that we give it the name m in parentheses, and we add some spacing between columns and rows to have space for the arrows. We set a fixed text depth and height to achieve proper baseline alignment, even though some letters have descenders and some don't.

This is our matrix code now:

```
\matrix (m) [matrix of nodes,
   row sep = 2em, column sep = 2em,
   nodes = {text depth = 1ex, text height = 2ex}
]
{
   tex & dvi & ps \\
       & pdf &    \\
};
```

In a matrix, the cell nodes have implicit names based on the matrix name, row, and column. Here, we set (m) as the matrix name, then we can address the node in the first row and second column as (m-1-2). It becomes clearer when we add all the arrows, so follow the numbering in this \draw command:

```
\draw [-stealth]
   (m-1-1) edge (m-1-2)
   (m-1-2) edge (m-1-3)
   (m-1-1) edge (m-2-2)
   (m-1-2) edge (m-2-2)
   (m-1-3) edge (m-2-2)
;
```

This code adds edges with `stealth` arrow tips between the chosen cells. The two commands give us the following output:

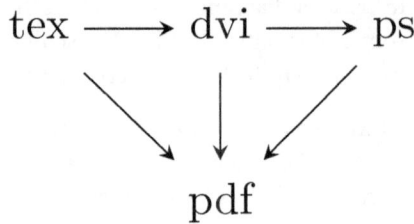

Figure 6.23 – A matrix diagram with arrows

If you don't like that naming scheme, you can choose your own cell node names. The vertical bars also work for choosing node names. When we write `|(d)|dvi` and `|(p)|pdf` in our matrix, we can later refer to them by the names `(d)` and `(p)`, like this:

```
\draw (d) -- (p);
```

That's useful when diagrams get bigger or when you insert additional cells that change the previous numbering, and it makes your code more readable to yourself.

Personally, I like that two-step approach:

1. Positioning all nodes with the `\matrix` command.
2. Connecting the nodes with arrows with the `\draw` command.

In *Chapter 14*, *Drawing Diagrams*, we will use predefined styles and generate diagrams in an automated way.

Summary

In this chapter, you learned about visualizing information in hierarchical structures such as trees, and you learned new short syntax elements for creating graphs and diagrams. Combine this with all you learned about nodes, edges, and styles, and you have become a master at drawing diagrams.

In the next chapter, we will learn more about advanced drawing techniques.

Further reading

The full reference for trees and graphs is the TikZ manual at `https://texdoc.org/pkg/tikz` in PDF format.

These sections are the most relevant for this chapter:

- *Part III, Section 21, Making Trees Grow*, is the reference for all basics. The direct online link is `https://tikz.dev/tikz-trees`. *Part V, Section 76, Tree Library*, describes the `tree` library in depth. The online manual link is `https://tikz.dev/library-trees`.

- *Part III, Section 19, Specifying Graphs*, is the `graphs` library reference and is available online at `https://tikz.dev/tikz-graphs`. *Part IV, Graph Drawing*, is a comprehensive part dedicated to algorithmic graph drawing, where TikZ computes the layout for you, available at `https://tikz.dev/gd`.

- *Part III, Section 20, Matrices and Alignment*, explains the basic matrix features; its link is `https://tikz.dev/tikz-matrices`. *Part V, Section 59, Matrix Library*, is the reference for the additional styles, such as `matrix of nodes`, and is found at `https://tikz.dev/library-matrix`.

Online TikZ galleries provide interesting examples with source code:

- `https://tikz.net/tag/trees` shows examples of trees in application in science
- `https://texample.net/tikz/examples/feature/trees` shows examples of TikZ trees
- `https://texample.net/tikz/examples/tag/mindmaps` shows some different examples of mind maps
- `https://texample.net/tikz/examples/feature/matrices` is a collection of matrix examples

The *LaTeX Cookbook* has further examples of trees and mind maps; you can browse them at `https://latex-cookbook.net` and read the full explanation in the book.

7
Filling, Clipping, and Shading

Now, we enter the book's second part and start with advanced drawing techniques.

The first chapters explained drawing paths, geometric objects, and path elements such as nodes and edges. This chapter will focus on areas enclosed by a path for filling and clipping.

Specifically, we will deal with the following:

- Filling an area
- Understanding a path interior
- Clipping a drawing
- Reverse clipping
- Shading an area

At the end of the chapter, you will be proficient in coloring and clipping and know two different ways to define areas with self-intersecting paths and unconnected path segments.

Technical requirements

You can work with the code at https://tikz.org/chapter-07 or use Overleaf apart from your local LaTeX setup with a complete TikZ package.

The code examples are available on GitHub at https://github.com/PacktPublishing/LaTeX-graphics-with-TikZ/tree/main/07-filling-clipping-shading.

In this chapter, we will use the shadings library.

We will use many colors in this chapter; the paper version of the book is grayscale. Therefore, it's highly recommended to read the e-book version in PDF or Kindle format or read the TikZ.org chapter page to see the images in their full glory.

Filling an area

We already used the `fill` option in previous chapters. Now, we will take a closer look at filling. Until now, we filled node shapes and simple geometric areas, for example, with `\node[fill]` and `\draw[fill]`. There are command aliases:

- `\fill` is equivalent to `\path[fill]`; we use it for filling without drawing a border
- `\filldraw` is equivalent to `\path[draw, fill]` and `\draw[fill]`; in this case, we will get a border

Both commands take options for colors, such as `\fill[yellow]` for yellow filling and `\filldraw[fill=yellow,draw=red]` for a yellow-filled area with a red border.

If a path encloses an area, TikZ closes it by connecting the last coordinate with the first coordinate, and then it fills it. Of course, it's good to close the path ourselves by ending with the same coordinate as the path start coordinate. We can do that with a short generic statement by adding `-- cycle` to the path code, which means connecting finally to the start point. We will do that in the following sections.

For filling node shapes and simple geometric figures, it's clear what TikZ must do. If the path is more complex, it may be challenging. For example, a path can consist of several lines, arcs, and circles; they may be connected, or they may not touch each other, and a path may self-intersect. TikZ then has to decide what the inside area is to be filled, and what the outside area is.

Suppose our path produces several areas, possibly one area containing another one, or two overlapping. In that case, there can be different opinions about the inside area to be filled and what is on the outside.

In the next section, we will investigate the interior of complex paths and apply our filling commands to work hands-on with them.

Understanding the path interior

Paths can be more complex than just a circle or convex polygon. Take a quick look at these three examples:

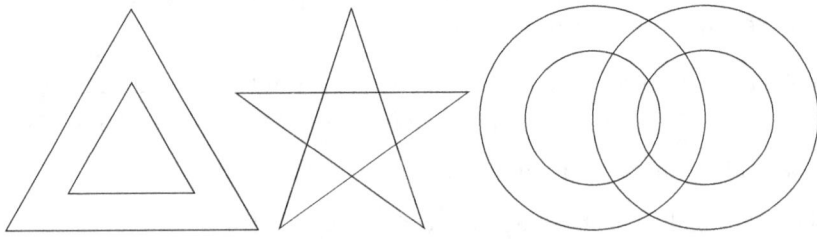

Figure 7.1 – Various paths

In *Figure 7.1*, we have different questions regarding how to color areas, such as the following:

- In the triangle, do we consider the inner triangle as inside or outside? Can we fill the complete big triangle or just the space between them?
- Can we color the full star shape or just the spikes?
- Can we color small segments of that multiple-circle path?

This section will help us to answer these questions. Regarding the third point, the following section, *Clipping a drawing*, will provide us with a tool.

For different area selecting and coloring, TikZ implements two different interior rules, used in computer graphics. We will discuss them both now.

The nonzero rule

Let's say we have a closed path. Naturally, that path has a direction from start to end. When am I inside or outside the path area? Imagine I'm standing at some point and then start walking straight toward the path, crossing it at least once:

- If I crossed the path and the path came from the left, and then I crossed the path again but now the path came from the right, I entered and left the inner area. This means I was outside.
- If I crossed the path just once and not again, I would be inside the path.

The following figures illustrate what I mean. Consider a triangle made with the following code:

```
\path (90:2) -- (210:2) -- (330:2) -- cycle;
```

To explain this, I've added markers indicating the direction and chosen two points to consider, as follows:

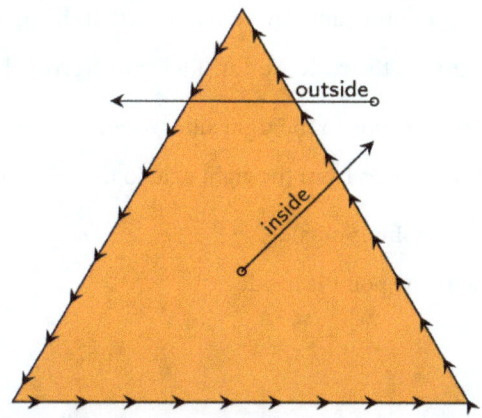

Figure 7.2 – The inside point and outside point

Now, we can enter and leave an inner area several times. Let's consider that situation with a path made with this code:

```
\path (150:1) -- (210:2)   -- (330:2)
   -- (30:1) -- (0,-0.5) -- cycle;
```

Again, we consider two points in this figure:

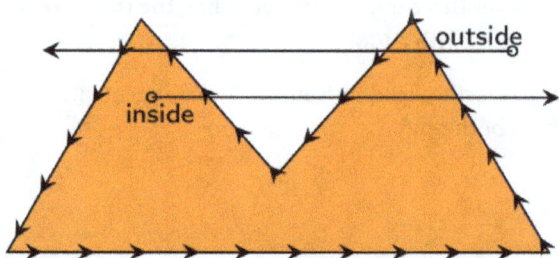

Figure 7.3 – The inside point and outside point with a complex path

When we consider the same approach shown in *Figure 7.2*, note that we have crossed the path several times. From the outside point, we have the same number of left and right crossings. From the inside point, we don't.

Such a walk can be described more formally – from our point of interest, we choose a ray toward infinity in any direction. Then, we count how many times that ray crosses the path segments, and we consider the direction as shown previously. We will track the number, starting with zero.

This leads us to the following approach:

- If our ray doesn't cross the path, we have a zero value. Because the path is closed, the ray would have to hit the path if the point is inside. So, zero means that the point is outside.
- Each time when the ray crosses the path, we consider the direction by how it meets the path:
  - If the path goes from the left side to the right side, we add one
  - We subtract one if the path goes from the right side to the left side
- If the final value is zero, the point is outside.
- If the final value is nonzero, the point is inside.

That's where the name **nonzero rule** comes from. Let's quickly check what we did before:

- In *Figure 7.2*, we calculated 1 − 1 = 0 for the outside point and -1 for the inside point
- In *Figure 7.3*, we had 1 − 1 + 1 − 1 = 0 for the outside point and -1 + 1 − 1 = -1 for the inside point

In both figures, the area where we get a nonzero value for any point is painted orange.

When we have several path segments, we can decide the direction of each path segment to influence the outcome. That gives us more control over the filling.

We will take this path as an example:

```
\path (90:2)  --  (210:2)  --  (330:2)  --  cycle
      (90:1)  --  (210:1)  --  (330:1)  --  cycle;
```

That gives us an outer triangle and a smaller inner triangle. They have the same direction, as we can see in this figure:

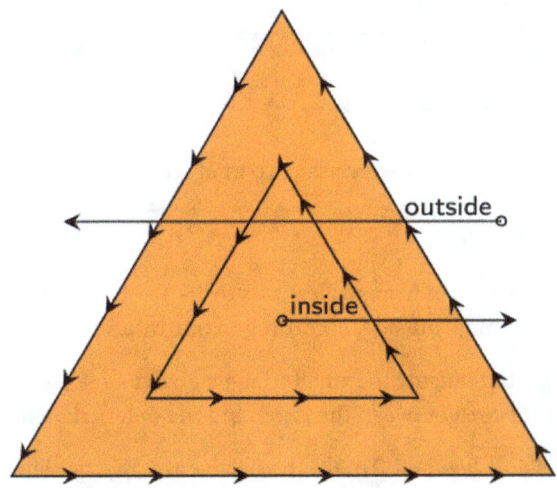

Figure 7.4 – Filling a path with two parts

With our `nonzero rule`, we calculate for the inside point: -1 − 1 = -2. It's nonzero. For the outside point, we get 1 + 1 − 1 − 1 = 0. The whole area of the big triangle is an inside area and will be filled.

What happens if we change the direction of the smaller triangle path? The code would now be the following:

```
\path (90:2) -- (210:2) -- (330:2) -- cycle
      (90:1) -- (330:1) -- (210:1) -- cycle;
```

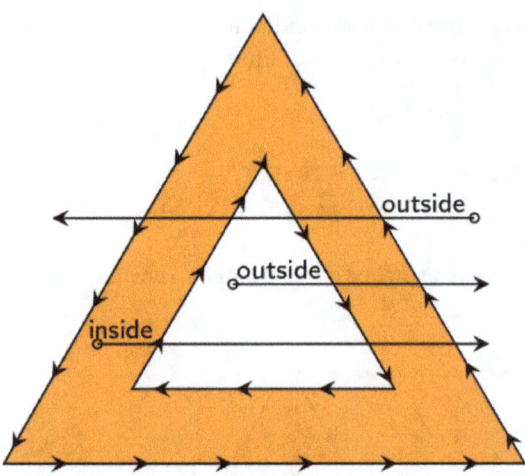

Figure 7.5 – Reversing a part of the path

We can do the same exercise as before.

Our first outside point, right of the big triangle, leads us to 1 - 1 + 1 - 1 = 0. A point inside the small triangle gives us 1 - 1 = 0. That's now outside, according to our rule.

A point inside the big triangle but outside the small triangle gives us 1 + 1 - 1 = 1 and is nonzero. So, our inner area to be filled is the area between the inner and the outer triangle.

Each approach in *Figure 7.4* and *Figure 7.5* makes perfect sense. By choosing the direction of path segments, we can choose the kind of filling that gives us some flexibility.

When we decide on the second approach, our code for filling the path would be the following:

```
\fill[orange]
    (90:2) -- (210:2) -- (330:2) -- cycle
    (90:1) -- (330:1) -- (210:1) -- cycle;
```

The result is this filled area, bordered by triangles:

Figure 7.6 – A filled area between triangles

`nonzero rule` can be chosen as an option, such as `\fill[orange, nonzero rule]`, but it's the default rule, so you don't need to choose it explicitly.

The other rule is not as complex – I promise.

The even odd rule

Again, to decide whether a point is inside or outside the path-bordered area, we will consider a ray from that point toward infinity.

This time, the method's surprisingly simple:

- We count how often the ray crosses the path
- If the total number is even, the point is outside
- If the number is odd, the point is inside

Because of this decision, we call this approach the `even odd rule`.

With this rule, we get the same result for the filled area in *Figure 7.2*, *Figure 7.3*, and *Figure 7.5*. Only *Figure 7.4* is not possible. For such a pair of triangles, we would always get *Figure 7.5* as a result. That's because it's independent of the direction of the path.

Since that rule is not set by default, we have to set it explicitly. The code for *Figure 7.6* would be as follows:

```
\fill[orange, even odd rule]
   (90:2) -- (210:2) -- (330:2) -- cycle
   (90:1) -- (330:1) -- (210:1) -- cycle;
```

Of course, we can set the option for a whole TikZ picture with the following:

```
\begin{tikzpicture}[even odd rule]
```

With this rule, any walk over the partial areas alternates between filled and non-filled. In other words, with any two adjacent partial areas, one will be filled and one won't. That's also a naturally expected result.

Comparing the nonzero rule and the even odd rule

Let's compare the filling rules to decide when to use which one.

The pros of `even odd rule` are as follows:

- It's easy to understand and verify in a drawing
- The result is the same, independent of path direction and drawing order
- We can simply add a segment to open a hole in an area

The pros of `nonzero rule` are as follows:

- It allows flexibility; we can change the outcome if we change the direction of a path segment
- We can have adjacent areas with the same filling
- We may use fewer path segments to fill an area because crossing path segments resulting in adjacent areas don't have to open outside areas within our main area

Let's explore the last point. We will look at a star path:

```
\path (90:1) -- (234:1) -- (18:1)
   -- (162:1) -- (306:1) -- cycle;
```

This is how the path looks:

Figure 7.7 – A star path

Now, we want to fill that star with light blue. With the default `nonzero rule`, we simply change `\path` to `\fill[blue!50]`. With `even odd rule`, we change `\path` to `\fill[blue!50, even odd rule]`.

In the following figure, we compare both results:

Figure 7.8 – Nonzero rule on the left and even odd rule on the right

We already know that adjacent areas cannot have the same color with `even odd rule`. Our way to solve it with `even odd rule` would be to construct the star using only non-crossing lines. Instead of only 5 coordinates, we would have to use 10 coordinates. You can draw this as an exercise, or look at it on the book's website or GitHub.

For such reasons and more flexibility, `nonzero rule` is the default rule. It's also the default filling rule in the **Scalable Vector Graphics (SVG)** format.

This section was about determining an area enclosed by a path, which we filled with a color. We can also use this area to restrict and cut out a part of the drawing. That is called clipping and is our next topic.

Clipping a drawing

Clipping means cutting pieces from a drawing or a path. In other words, it means restricting a picture to a specific area, called the **clipping area** or **clipping path**. The clipping area can be a rectangle, a circle, or an arbitrary path.

First, an easy example. Let's cut the corners of the filled triangle from *Figure 7.6*. A circle like this will clip it:

Figure 7.9 – A circle for clipping a triangle

First, we define the clipping path:

```
\clip (0,0) circle (1.5);
```

Then, we proceed with our drawing:

```
\fill[orange] (90:2) -- (210:2) -- (330:2) -- cycle
              (90:1) -- (330:1) -- (210:1) -- cycle;
```

The result, as expected, is the following image:

Figure 7.10 – A clipped triangle

Similar to `fill`, we can use `clip` as a command and an option:

- `\clip` is equivalent to `\path[clip]`; we can use it to declare a clipping path without drawing anything
- `\draw[clip]` is equal to `\path[draw, clip]` and will both draw the path and define it as a clipping path

We can specify several clipping paths that add up, meaning that all of them restrict the following drawing (*Figure 7.11*), determined by their intersection.

When we want to stop the clipping and draw more elements in the picture that will not be clipped, we can use a scope for clipping, such as the following:

```
\begin{scope}
   \clip (0,0) circle (1.5);
   \draw ...
\end{scope}
\draw ...
```

When the scope ends, the clipping restriction is lifted.

The clipping area is the **interior** of the clipping path. As we saw in the previous section, the interior can be selected using different rules. The `\clip` command does not accept interior rule options; however, we can specify the desired interior rule by using it as an option to a `scope` environment around `\clip`, like in the previous code snippet, plus the following option:

```
\begin{scope}[even odd rule]
```

We can also set the option for the whole TikZ picture; we don't need a `scope` environment. That's done as usual:

```
\begin{tikzpicture}[even odd rule]
```

Let's apply this to our multiple-circle example from *Figure 7.1*. Our challenge is to fill the small segment depicted in the following figure:

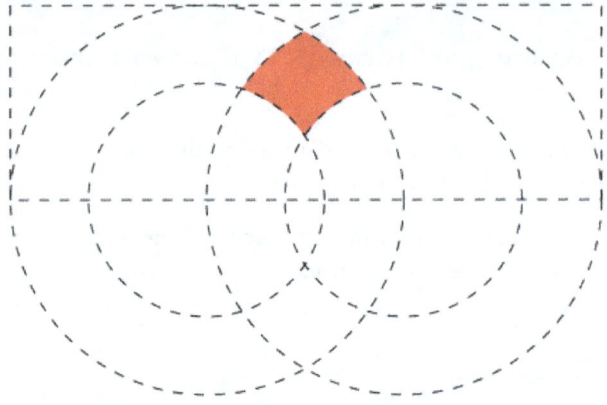

Figure 7.11 – A segment of intersecting rings

We see two rings here. We can use the left ring as a clipping path and draw the right one. Additionally, we will use the rectangle in the picture to restrict to the upper half and not color the similar small segment at the bottom of the figure.

We can draw the filled rings this way:

```
\fill[red!70]   (-1,0) circle (1.2)  (-1,0) circle (2);
\fill[red!70]   (1,0) circle (1.2)   (1,0) circle (2);
```

This results in this output:

Figure 7.12 – Filled rings with nonzero rule

Unfortunately, that's far from what we want; the default `nonzero` rule simply fills the whole area enclosed by the rings. We will switch to `even odd rule` for the whole picture:

```
\begin{tikzpicture}[even odd rule]
  \fill[red!70] (-1,0) circle (1.2) (-1,0) circle (2);
  \fill[red!70]  (1,0) circle (1.2)  (1,0) circle (2);
\end{tikzpicture}
```

When we compile, the output comes closer to our goal:

Figure 7.13 – Filled rings

We can now clip it with a rectangle by inserting this line before our two `\fill` commands:

```
\clip (-3,0) rectangle (3,2);
```

The area is clipped to the top part. To see the clipping path, I just added the rectangle with dashed lines here:

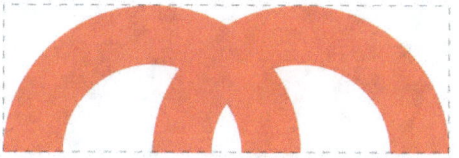

Figure 7.14 – Clipped rings

Now, instead of filling both rings, let's use the left ring as a clipping path and only fill the right ring. Our full code for the picture is now as follows:

```
\begin{tikzpicture}[even odd rule]
  \clip (-3,0) rectangle (3,2);
  \clip (-1,0) circle (1.2) (-1,0) circle (2);
```

```
    \fill[red!70] (1,0) circle (1.2) (1,0) circle (2);
\end{tikzpicture}
```

The final and fully clipped result is shown in the following figure:

Figure 7.15 – A clipped segment of the rings

Refer to *Figure 7.11* to verify that we colored the right area. That clipping also used the `even odd rule` to determine the clipping area.

The following section will show how to color other areas of our intersecting rings.

Reverse clipping

Sometimes, we may want the opposite of clipping – instead of cutting away everything that's not in our clipping area, we would only cut out that part inside a particular area.

First, let's look at regular clipping again. This time, we will color a different segment of our pair of intersecting rings from *Figure 7.11*, as shown in the following figure:

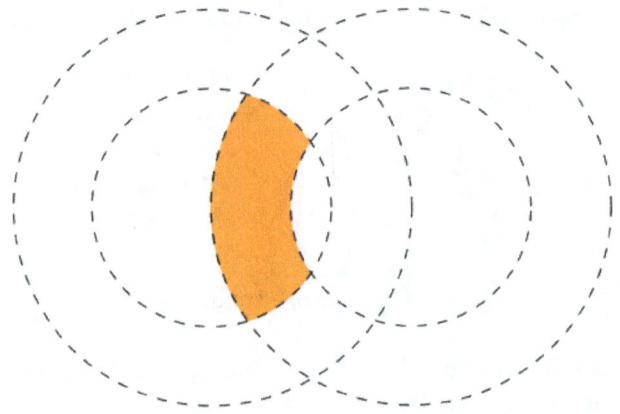

Figure 7.16 – Another segment to fill

The code is a straightforward exercise of what we already know. We will choose the smaller circle on the left as a clipping path and draw the right ring. The code is as follows:

```
\begin{tikzpicture}[even odd rule]
  \clip (-1,0) circle (1.2);
  \fill[orange] (1,0) circle (1.2) (1,0) circle (2);
\end{tikzpicture}
```

Now to our challenge, coloring the other part of the right ring, as follows:

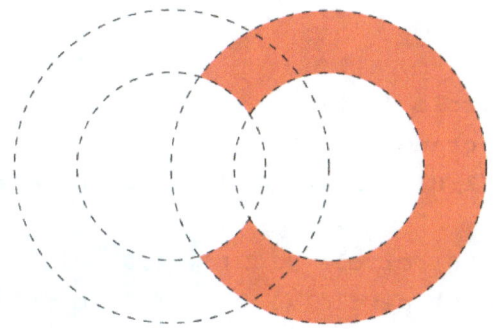

Figure 7.17 – Another segment to fill

To also practice limiting the clipping effect by scopes, we will use clipping, fill the segments in the same picture, and draw our rings with dashed lines.

The point of our solution is not to clip with the small circle but to choose the circle with an area around it that's big enough to cover the rest of the drawing, as the clipping path. So, we will choose a big rectangle together with a small circle as our clipping area, resulting in a new clipping area, as follows:

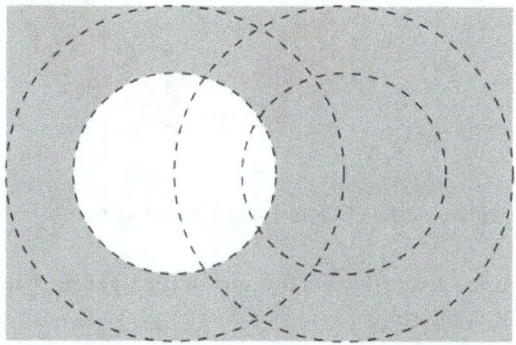

Figure 7.18 – The reverse clipping area

The new clipping area in gray is like the inverse of the small left circle, like everything except it. When we clip the area marked in gray here, with the right ring, we get the right piece cut out, and the rest of the ring is colored.

The complete code of our exercise is the following, where the reverse clipping part is highlighted:

```
\begin{tikzpicture}[even odd rule]
  \begin{scope}
    \clip (-1,0) circle (1.2);
    \fill[orange] (1,0) circle (1.2) (1,0) circle (2);
  \end{scope}
  \begin{scope}
    \clip (-1,0) circle (1.2)
          (-2,-2) rectangle (3,2);
    \fill[red!70] (1,0) circle (1.2) (1,0) circle (2);
  \end{scope}
  \draw[dashed] (-1,0) circle (1.2) (-1,0) circle (2);
  \draw[dashed]  (1,0) circle (1.2)  (1,0) circle (2);
\end{tikzpicture}
```

This code gives us the following:

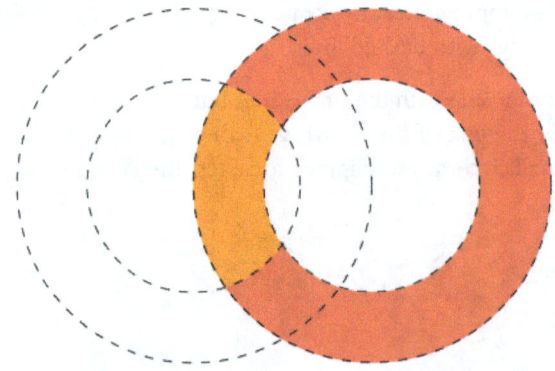

Figure 7.19 – Colored segments of a ring

The red area is the reverse clipped area, which is the part of the right ring that is not in the left small circle. So, we cut the circle out of the ring.

We used a rectangle of well-matching size for the reverse clipping path. If we are not sure about the final dimensions of the drawing, we could choose a giant rectangle to ensure we cover everything. However, the clipping path is also considered in the bounding box calculation. If we make it too big, we could get undesired white space. The solution here is to add the `overlay` option to `\clip` so that this path is ignored in the bounding box dimensions. In the previous example, we could also write the following:

```
\clip[overlay] (-1,0) circle (1.2)
            (-20,-20) rectangle (30,20);
```

You can read more about the `overlay` option and layers in general in *Chapter 9, Using Layers, Overlays, and Transparency*.

Until now, we have always used a uniform color to fill an area. In the next section, we will turn to more fancy filling.

Shading an area

Instead of filling an area with a single color, we can use several colors with a transition between them. TikZ provides several kinds of smooth transitions in different ways with the `shade` action.

Similar to `fill`, `shade` can be used as a command and an option:

- `\shade` is equivalent to `\path[shade]`; use it for shading without drawing a border
- `\shadedraw` is the same as `\path[draw, shade]` and `\draw[shade]`; it produces the shading and adds a border

We will look at several shading styles in the following few sections. The first three, `axis`, `radial`, and `ball`, are included with TikZ by default. To use the other shadings, load the corresponding library in your preamble with `\usetikzlibrary{shadings}`.

Often, you don't need to choose a style explicitly. Depending on your color options, TikZ can automatically determine one, so it's pretty intuitive.

We will look at examples with some randomly chosen colors.

Axis shading

Axis shading is linear shading between top and bottom, left and right, or at a chosen angle.

The following command draws a square with a red color at `top` that gradually transitions to yellow at `bottom`:

```
\shadedraw [top color=red, bottom color=yellow]
   (0,0) rectangle (1,1);
```

This command draws a square with a red color on `left` that gradually transitions to yellow on `right`:

```
\shadedraw [left color=red, right color=yellow]
   (1.5,0) rectangle (2.5,1);
```

The output can be seen here:

Figure 7.20 – Squares with shading

Now, try a shading angle of 30 degrees counterclockwise, applied to our previous triangle example:

```
\shade[top color=red, bottom color=yellow,
   shading angle=30]
   (90:2) -- (210:2) -- (330:2) -- cycle
   (90:1) -- (330:1) -- (210:1) -- cycle;
```

This produces the following:

Figure 7.21 – A triangle with rotated shading

This also shows and means that we can apply shading in the same way and with the same interior rules as filling.

We can use it for a 3D-like effect in a drawing. For example, here, we quickly shade three areas to simulate some light:

```
\shade[left color=black!60, right color=black!10]
   (0,0,0) -- (1,0,0)  -- (1,1,0)  -- (0,1,0);
\shade[left color=black!10, right color=black!80]
   (1,0,0) -- (1,0,-1) -- (1,1,-1) -- (1,1,0);
\shade[bottom color=black!10, top color=black!80]
   (0,1,0) -- (0,1,-1) -- (1,1,-1) -- (1,1,0);
```

This gives us a cube-like appearance:

Figure 7.22 – A shaded cube

Using a rectangle with a rounded corner shape and top-to-bottom shading, we can easily create a 3D-button style for diagrams.

We can optionally specify a middle color. Because the color in the middle is automatically interpolated and we want to override it, we would have to set it after the other colors. Then, a smooth transition is made between these three colors, as shown here, with squares again:

```
\shadedraw [left color= black, right color=red,
   middle color=white] (0,0) rectangle (1,1);
\shadedraw [bottom color=black, top color=blue,
   middle color=orange] (1.5,0) rectangle (2.5,1);
```

Those two commands produce these filled squares:

Figure 7.23 – Axis shading with a middle color

When we choose the colors for the left, right, top, bottom, or middle, the `shading=axis` style is implicitly chosen. We can set it, but we don't have to.

TikZ uses such shading for the color transition between concepts in the mind maps we saw in the previous chapter.

Radial shading

We can add shading from the inside to the outside of an area, like this, best shown in a circle:

```
\shade[inner color=yellow, outer color=red]
   (0,0) circle (1);
```

That gives us this circle:

Figure 7.24 – Radial shading

Again, `shading=radial` is implicitly set, as TikZ recognizes inner and outer colors.

If you want that "highlight" effect moved away from the center, you could shade a differently sized circle and clip it to the desired circle.

Ball shading

When you set `ball color`, TikZ applies `shading=ball` and transitions between the dark and the light color in a highlight spot, just as if some light was shining on a ball. That results in a 3D effect.

Here, we produce red, green, and blue balls:

```
\shade[ball color=red]    (0,0)   circle (1);
\shade[ball color=green]  (2.5,0) circle (1);
\shade[ball color=blue]   (5,0)   circle (1);
```

The output is the following:

Figure 7.25 – Ball shadings from left to right – red, green, and blue

We previously used ball shading for the head and eyes in the smiley drawing in the second chapter of this book.

All of the following shading types require `\usetikzlibrary{shadings}`.

Bilinear interpolation

In computer graphics, the bilinear interpolation of colors uses the average of the four nearest pixels' color values to determine pixel color. So, you can have a rectangle with a chosen color at each corner, and then every point in the area is interpolated between all four.

It's good to see it as an example. Here, we will define the corner color values, and TikZ does the interpolation:

```
\shade[upper left=green, upper right=blue,
       lower left=red,   lower right=yellow]
      (0,0) rectangle (1,1);
```

Look at the output:

Figure 7.26 – Bilinear interpolation shading

Again, you can use a larger area for the shading and clip it to your area of interest to get a specific effect.

Color wheel

This is a predefined circular shading. Let's have a quick chat about color models before we proceed.

In computer graphics, we have different color models. One is CMYK, composed of four base colors – cyan, magenta, yellow, and key (which stands for black). It's often used for printing. It starts on white paper, and every added color reduces the reflected light. That's why it's called a subtractive color model. When all colors are added with full saturation, we get black.

Another one is RGB (which stands for red, green, and blue), where the color is made with light, like on a monitor. More light means brighter colors, and stronger colors also mean more light. With maximum strength, the base colors red, green, and blue add up to white. It's an additive model.

That's good to know when we look at the following color wheel. This shading produces an RGB color wheel:

```
\shade[shading=color wheel] (0,0) circle (1);
```

Without much ado, this is the output:

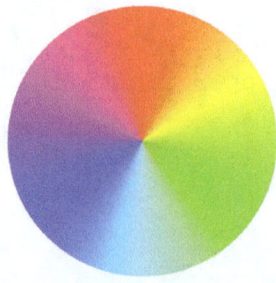

Figure 7.27 – Color wheel shading

The explanation beforehand was to understand the lightness between the color slices.

We get an RGB color ring when we apply the color wheel shading to our previous ring example:

```
\shade[shading=color wheel, even odd rule]
    (1,0) circle (1.2) (1,0) circle (2);
```

We get this picture:

Figure 7.28 – The color ring

It doesn't have to be circular. Note that we have three base colors, and applying them to our example triangle is just natural:

```
\shade[shading=color wheel]
    (90:2) -- (210:2) -- (330:2);
```

Our color triangle is this:

Figure 7.29 – The color triangle

There are no more options here. However, there are two variations of this shading. The first one is with a black center instead of white:

```
\shade[shading=color wheel black center] (0,0) circle (1);
```

This gives us the following:

Figure 7.30 – The color wheel with a black center

This is with the white center:

```
\shade[shading=color wheel white center] (0,0) circle (1);
```

And we get the following:

Figure 7.31 – The color wheel with the white center

We can choose what fits our purpose.

Summary

This chapter was challenging, but now that you have completed it, you know how to define and restrict areas for filling in your drawings with a color or interpolated colors.

In the next chapter, our drawings will not get more accessible but more advanced, and they will also get fancier.

Further reading

In the TikZ manual, available at `https://texdoc.org/pkg/tikz` in PDF format, these sections are relevant:

- *Part III, Section 15, Actions on Paths*, is about filling, clipping, and shading, and it explains the interior rules concisely. The direct online link is `https://tikz.dev/tikz-actions`.
- *Part V, Section 69, Shadings Library*, is the reference for always-defined and additional shading types, and it can be found at `https://tikz.dev/library-shadings`. *Part IX, Section 114, Shadings*, explains the backend programming behind shadings and can help you to create your own shading, such as tuning a ball shading with different highlighting. It's available online at `https://tikz.dev/base-shadings`.
- The web page `https://www.w3.org/TR/SVG/painting.html#WindingRule` by the **World Wide Web Consortium** (**W3C**) describes the nonzero and even odd interior rules from the point of view of the SVG standard.
- The TikZ galleries at `https://tikz.net` and `https://texample.net/tikz/examples/` contain many examples, which you can find with the `fading`, `clipping`, and `shading` keywords.

8
Decorating Paths

So far, we have created our drawings with paths consisting of straight or bent lines, various shapes, and text. TikZ offers us tools to make paths fancier. With a few options, we can change the path so that it has a bumpy or wavy appearance or one with a zigzag line. We can add markers such as arrows to paths and let the text flow along paths.

In this chapter, we will work on the following topics:

- Pre- and post-actions for using a path multiple times
- Understanding decorations
- Variations of decoration commands, options, and libraries
- Exploring the available decoration types
- Nesting decorations
- Adjusting decorations

Once you've learned the tools of this chapter, you will be able to add fancy details to your drawings and know how to add braces, markers, and curvy text to scientific illustrations.

Technical requirements

All the code examples are available at `https://tikz.org/chapter-08`. You can download the code from GitHub at `https://github.com/PacktPublishing/LaTeX-graphics-with-TikZ/tree/main/08-decorating-paths`.

In this chapter, we will use the `decorations` library. Specifically, we will load the `decorations.pathmorphing`, `decorations.pathreplacing`, `decorations.text`, `decorations.markings`, `decorations.shapes`, and `decorations.fractals` sub-libraries.

We will see a lot of examples in this chapter. Therefore, we will often show just code snippets or single commands. To see complete examples and to run the code, you can visit this chapter's page at `TikZ.org`.

Pre- and post-actions for using a path multiple times

We can have an arbitrarily complex path we want to use several times. TikZ provides an efficient way that spares us repeating coordinates.

Let's look at the following path from *Figure 7.4* from the previous chapter:

```
\draw[orange, line width=3mm]
    (90:2) -- (210:2) -- (330:2) -- cycle;
```

This draws an orange triangle:

Figure 8.1 – A simple triangle path

Let's say we want to draw this path several times with different colors and line widths to get a nice striped effect. The straightforward way to do this is by repeating the \draw command, like so:

```
\draw[red, line width=5mm]
    (90:2) -- (210:2) -- (330:2) -- cycle;
\draw[orange, line width=3mm]
    (90:2) -- (210:2) -- (330:2) -- cycle;
\draw[yellow, line width=1mm]
    (90:2) -- (210:2) -- (330:2) -- cycle;
```

Those three commands give us the following triangle in a mix of red, orange, and yellow colors:

Figure 8.2 – A repeated path with multiple colors and different widths

TikZ allows us to specify additional actions on paths as options that should happen before or after the path is drawn and applied to the same path. We can put the additional actions and options into a `preaction` option and a `postaction` option, as follows:

```
\draw[orange, line width=3mm,
   preaction  = {draw, red,    line width=5mm},
   postaction = {draw, yellow, line width=1mm}]
   (90:2) -- (210:2) -- (330:2) -- cycle;
```

This draws the path first in red, then in orange, and finally in yellow, with different widths, resulting in this overlapping effect. It gives us the same output as *Figure 8.2* and is significantly shorter.

We can use multiple pre- and post-actions; they will be applied in their order of appearance in the code.

In this chapter, we will create decorations based on paths added to the path, so it's natural to use them as a `preaction` or `postaction` option. Therefore, we will use these options in the following sections.

Understanding decorations

The TikZ syntax can be pretty verbose. That's particularly the case with decorations. Even adding arrow tips along a path can be wordy. As this is an excellent example of what TikZ decorations can do for us, let's try this.

Let's create an arrow from (0,0) to (2,0). We can do this with the following code:

```
\draw[-stealth] (0,0) -- (2,0);
```

We have an arrow tip at the end of the path, but we also want to have arrow tips along the way. First, we must load the `decorations.markings` library:

```
\usetikzlibrary{decorations.markings}
```

Then, we must choose the `decorate` option as `postaction`, with a `decoration` type of `markings`, in steps of 0.2 between positions 0.2 and 1:

```
\draw[-stealth, postaction = decorate,
   decoration = {markings,
   mark = between positions 0.2 and 1 step 0.2
   with {\arrow{stealth}}}]
      (0,0) -- (2,0);
```

The syntax will be explained in detail later in this chapter in the *Adding markings* section. For now, we want to see the effect of `decoration`; this is displayed in the following figure:

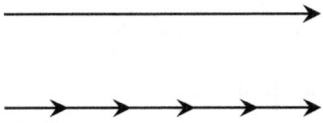

Figure 8.3 – An arrow with and without decoration

We can use an arbitrary path like this:

```
\draw[-stealth, postaction=decorate,
   decoration = {markings, mark = between positions 0.1
   and 1 step 0.1 with {\arrow{stealth}}}]
      (0,0) arc(180:0:1) arc(-180:0:1);
```

This gives us arrows following the arcs:

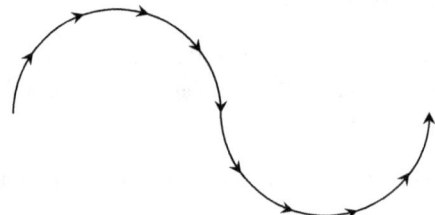

Figure 8.4 – Arrowhead markings along a curvy path

Understanding decorations

Decorations are often additions to a path but can also modify or replace a path.

There are different ways of declaring a decoration. Let's choose a simple zigzag path and see how we can state it.

In *Figure 4.1*, we had a straight arrow between two nodes, (tex) and (pdf). We will decorate it so that we can change it. Two things are needed:

- We have to state the `decorate` option.
- We need to define the `decoration` type. It can contain additional option values; in that case, we must enclose it in braces. Later examples will show this.

Note that we won't use `preaction` or `postaction` because we don't want to overwrite the path; instead, we want to change it. TikZ calls this **path morphing**.

The basic syntax for decorating the path looks like this:

```
\draw[decorate, decoration=zigzag, ->] (tex) -- (pdf);
```

Here, the decoration applies to the whole path from start to end. It gives us the following figure, which you can compare with *Figure 4.1*:

Figure 8.5 – A zigzag arrow

The second way is to use `decorate` as an operation inside the path. It allows you to restrict a decoration to just a part of the path, enclosed in curly braces:

```
\draw[->] decorate[decoration=zigzag] {(tex) -- (pdf)};
```

Furthermore, the `decorate` operation can add multiple decorations to a single path. We will use this in the last section, *Adjusting decorations*.

If you use only one decoration type in the TikZ picture and possibly several times, you can add the `decoration` type as an option to the whole picture, like this:

```
\begin{tikzpicture}[decoration=zigzag]
  \draw[decorate, ->] (tex) -- (pdf);
\end{tikzpicture}
```

This makes sense if a decoration has some verbose options, as in *Figure 8.3*.

We can also apply these options to an edge. Remember that an edge is a separate path with its own options. That's done like this:

```
\draw (tex) edge[->, decorate, decoration=zigzag] (pdf);
```

In the next section, we will see what decorations are available.

Exploring the available decoration types

TikZ has decorations that change a path, also called **morphing** a path. This will be our next topic.

Other decorations replace a path with symbols; we will see them later in this section.

To summarize and compare, we will provide some figures containing several drawings or paths. The figure caption will include the options used for each path, always in order from top to bottom. You can find the code for each figure on GitHub and on `TikZ.org` on the page for this chapter.

Morphing paths

Morphing means modifying a path to become, for example, a zigzag or jagged line. We will distinguish between linear and curvy morphing. We have to load the corresponding library in the preamble by using `\usetikzlibrary{decorations.pathmorphing}`.

The decorations of this library have optional values, such as these:

- `amplitude`: Determines how much the changed path goes above and below the original path
- `segment length`: The length of such a decoration cycle, going up and down

When we use decoration options, we have to enclose them in curly braces, like so:

```
decoration={zigzag, amplitude=2mm, segment length=3mm}
```

Some decorations have additional options, which will be mentioned shortly.

The following linear decorations are predefined:

- `zigzag`: Generates a zigzag line.
- `saw`: Creates a line that looks like a saw blade.
- `random steps`: Here, in each cycle, the move in the *X* and *Y* direction is randomly taken between -`amplitude` and +`amplitude` values.
- `lineto`: Replaces the path with straight lines. This makes sense if the original path is curvy.

This is what you get when you apply the decorations to a line:

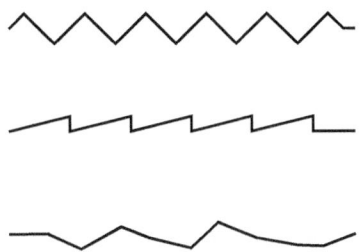

Figure 8.6 – Linear decorations on a line – zigzag, saw, and random steps

And this is how it looks applied to an arc:

Figure 8.7 – Linear decorations on an arc – zigzag, saw, random steps, and lineto

There are also curvy decorations:

- bumps: The path is replaced by half ellipses with segment length diameter and amplitude height.
- coil: This is like a spring along the path. The radius is amplitude. segment length is the width of a curl. Another option, aspect, defines the viewing angle: 0 is from the side, while 0.5 is more from the front, which is the default.

- `snake`: While looking like a snake from above, this is a sine curve with `segment length` as the wavelength and `amplitude` as the height interval.
- `bent`: Here, `amplitude` is how far it goes from a straight line, and `aspect` is how tight it bends. Try it out.

This is how it looks on a straight path:

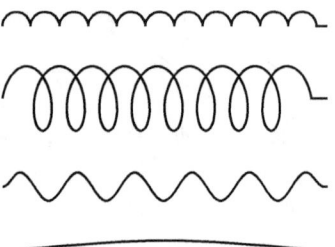

Figure 8.8 – Curvy decorations on a line – bumps, coil, snake, and bent

And this is how it looks applied to an arc:

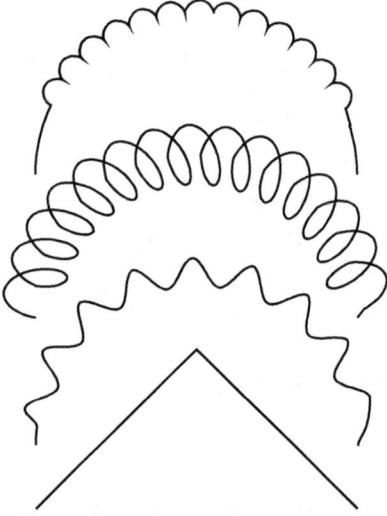

Figure 8.9 – Curvy decorations on an arc – bumps, coil, snake, and lineto

From morphing, we will now turn to replacing.

Replacing paths with ticks

We can replace a path with ticks, parentheses, or braces. Load the corresponding library by using `\usetikzlibrary{decorations.pathreplacing}`. The following replacements will become available:

- `border`: This indicates the border of an area, such as a wall, and is useful in mechanical engineering or architecture drawings. `segment length` is the distance between ticks, `amplitude` is the height of the ticks, and `angle` is the angle between the ticks and the path.
- `waves`: We get arcs along the path, in a distance of `segment length`, with a `radius` value you can specify and an `angle` value that is the opening between the beginning and the end of the path.
- `expanding waves`: This is like `waves`, but the circumference of the arcs will go from short to long.
- `ticks`: Here, you have orthogonal lines with a `segment length` distance. `amplitude` is how far it goes above and below the path.
- `brace`: Here, `amplitude` means how much the brace raises, and `aspect` is the fraction of the curvy middle path.

All of these options can be seen here:

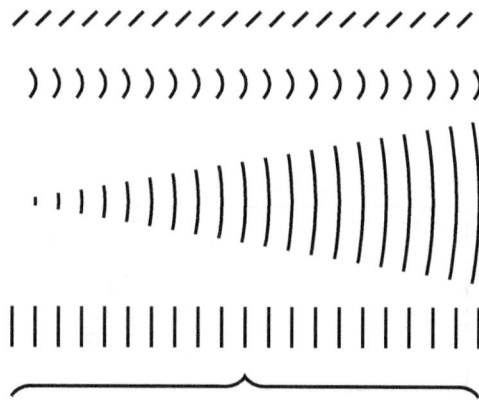

Figure 8.10 – Path-replacing decorations – border, waves, expanding waves, ticks, and brace

If we want to add such a decoration but keep the original path, we can use `preaction` or `postaction`, depending on whether we want to draw them behind or over the path. For example, we can add ticks to a line like this:

```
\draw[postaction = {draw, decorate,
    decoration = {ticks, segment length=1mm}}]
    (0,0) -- (2,0);
```

Figure 8.11 – A ticks decoration as a post-action

Braces are beneficial for showing widths and heights or summarizing anything. Here, you can see them being used to show measurements:

```
\begin{tikzpicture}[decoration=brace, font=\sffamily\tiny]
    \draw (0,0) rectangle (2,1);
    \draw[decorate]
        (0,1.05) -- node[above] {2 cm} (2,1.05);
    \draw[decorate]
        (2.05,1) -- node[above, sloped] {1 cm} (2.05,0);
\end{tikzpicture}
```

The output is as follows:

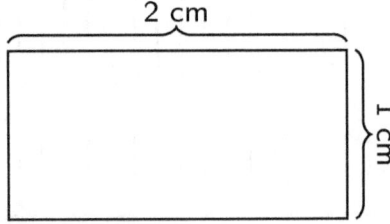

Figure 8.12 – Brace decorations for rectangle sides

With borders, ticks, braces, and node annotations, we can already make some technical drawings.

Decorating paths with text

TikZ has a text decorations library that you can load by using \usetikzlibrary{decorations.text}. We can use this library to replace paths with text following the same lines and curves. This is an example of how to write text along an arc:

```
\draw[decorate, decoration = {text along path,
   text = {text follows the path}}] (0,0) arc(180:0:1);
```

This command gives us the following output:

Figure 8.13 – Text along a curvy path

We can even let the text follow a path over multiple segments:

```
\draw[decorate, decoration = {text along path,
   text = {This is a long text along a path}}]
   (0,0) -- (1,0) arc(150:30:1.4) -- (5,0);
```

The text now runs like this:

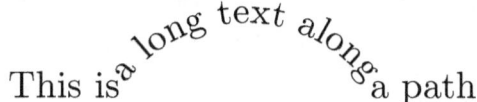

Figure 8.14 – Text along a path with multiple segments

As you can see, the path itself is not drawn; it's completely ignored. If we want, we can draw it with preaction or postaction.

Adding markings

We can annotate paths with markings, such as in *Figure 8.3*, where we added arrows. We can use any available TikZ arrow in the same way. First, we must load the library by using \usetikzlibrary{decorations.markings}.

Then, for example, this code draws triangle arrows along a path:

```
\draw[decorate,
  decoration = {markings, mark = between positions 0
  and 1 step 0.1 with {\arrow{Triangle}}}]
    (0,0) arc(120:60:1) arc(-120:-60:1);
```

You can choose any arrow you saw in *Chapter 4* as the \arrow parameter. Here is how it looks:

Figure 8.15 – Arrows along a path – stealth, triangle, and LaTeX[open]

If you load the decorations.shapes library and the shapes library itself, which we covered in *Chapter 3*, you can use any shape from that library in this way:

```
\draw[decorate, decoration = {shape backgrounds,
  shape=star, shape size=2mm}]
  (0,0) arc(120:60:1) arc(-120:-60:1);
```

Here are a few shapes with this path:

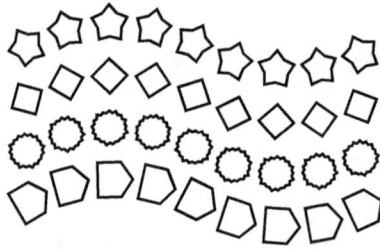

Figure 8.16 – Shapes along a path – star, diamond, starburst, and signal

There are options for width, height, distance, scaling, and rotation, which are explained in the TikZ manual. However, covering these here is outside the scope of this book.

Instead of the \arrow command for *Figure 8.15*, we can use any drawing command that will be repeated along the path, including nodes. Furthermore, we can use \pic commands. We are close to the end of this chapter, so let's have some fun and add the smiley from *Chapter 5*, *Figure 5.6*, where we defined \pic{smiley}. Since the smiley is bigger, we use a larger radius for the arc:

```
\draw[decorate, decoration = {markings,
  mark = between positions 0 and 1 step 0.04
  with {\pic {smiley};}}]
  (0,0) arc(120:60:40) arc(-120:-60:40);
```

This gives us the following figure:

Figure 8.17 – Smileys along a path

This example shows that TikZ is incredibly flexible and allows far more than just default decorations.

Adjusting decorations

If a decoration doesn't fit perfectly, there are options to adjust it. You can raise it by a positive or negative dimension using the `raise` option, such as in `decoration = {brace, raise=5pt}`.

The `mirror` option mirrors the decoration along the path. For example, `decoration = {brace, mirror}` would put the brace on the other side of the path.

We can also start a decoration later or end it earlier. These options keep a straight line of 5mm until the decoration starts:

```
pre=lineto, pre length = 5mm
```

If you have a curvy path, such as with arcs, you may prefer this:

```
pre=curveto, pre length = 5mm
```

Using the post option instead of pre, and post length, will do the same for the end of the path.

If you look closely at *Figure 8.7* and *Figure 8.9*, you will see that I used such pre and post options for the path at the top so that it looks better and more symmetric. You can see this in the code on this chapter's page at TikZ.org. This is the bump example from *Figure 8.9*, which shows how these options are used:

```
\draw[decorate, decoration={bumps,
    pre =curveto, pre  length=4.5mm,
    post=curveto, post length=3mm}]
      (0,0) arc(180:0:1);
```

You can use the pre and post options to start and end with different decorations, but the usual case is using lineto or curveto to let the decoration span a smaller part of the path.

Another way to combine different decorations is to use the decorate path operation, which was described after we presented *Figure 8.5*, to restrict decorations to a part of a path. For example, the following command draws an arrow with three different decorations in a row on a single path:

```
\draw[->] (0,0)
    decorate[decoration=bumps]  { -- (1,0) }
    decorate[decoration=zigzag] { -- (2,0) }
    decorate[decoration=saw]    { -- (3,0) };
```

Note how we used braces to indicate the scope of each decoration. That bumpy zigzag sawtooth-shaped arrow looks like this:

Figure 8.18 – Multiple decorations on a path

Decorations can be nested. The **Koch snowflake**, a basic fractal curve, is a good example to test this. Let's load the fractal decoration library by using \usetikzlibrary{decorations.fractals}. Now, we have a decoration with that name; we can add it to our \draw commands as usual or to our picture for simpler \draw commands, as follows:

```
\begin{tikzpicture}[decoration=Koch snowflake]
```

This command will draw a straight line from the origin (0,0) to the right with a 1 cm length:

```
\draw (0,0) -- (3,0);
```

Now, we can change that line by adding the decoration:

```
\draw decorate{ (0,0) -- (3,0) };
```

The path has been changed to a zigzag, as follows:

Figure 8.19 – The Koch snowflake decoration

Now, what happens if we decorate the path from *Figure 8.19* again with the same decoration?

```
\draw decorate{decorate{ (0,0) -- (3,0) }};
```

The same zigzag curve will now replace each line segment:

Figure 8.20 – The Koch snowflake decoration iterated

We can do the same again:

```
\draw decorate{decorate{decorate{ (0,0) -- (3,0) }}};
```

The shape gets finer:

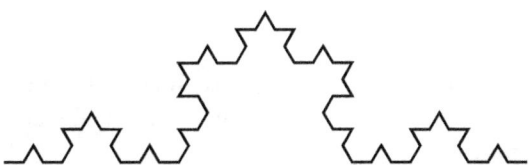

Figure 8.21 – The Koch curve after three iterations

If this were repeated again and again, indefinitely, the result would be the so-called **Koch curve**, named after the mathematician Koch, who originally described it.

But why is it called a snowflake? We can immediately see this when we apply the decoration with many iterations on a triangle:

```
\draw decorate{decorate{decorate{decorate{decorate{
    (210:2) -- (90:2) -- (330:2) -- cycle}}}}};
```

The output now looks like a fine snowflake:

Figure 8.22 – The Koch snowflake

The `decorations.fractals` library contains three more curves as decorations, called `Koch curve type 1`, `Koch curve type 2`, and `Cantor set`, which all work similarly: they replace a straight line with line segments. Try them and nest them.

Summary

In this chapter, you learned about repeated effects on paths. Apart from adding decorative effects, you can use ticks, braces, and arrows in scientific drawings, such as for mathematics or physics, or zigzags and coils in mechanical engineering.

You also learned how to work with pre- and post-actions on paths to draw something before or after a particular path. In the next chapter, we will explore such a concept for the whole picture: we will draw on the background and the foreground of pictures. Furthermore, we will use transparency to be able to see what is in the background.

Further reading

The TikZ manual at `https://texdoc.org/pkg/tikz` provides additional information on the topics that were covered in this chapter:

- *Part III, Section 15.10, Doing Multiple Actions on a Path*; this section is about `preaction` and `postaction`. It's available online at `https://tikz.dev/tikz-actions#sec-15.10`.
- *Part III, Section 24, Decorated Paths*, is comprehensive documentation on decorations. It's available online at `https://tikz.dev/tikz-decorations`.
- *Part V, Section 50, Decoration Library*, is the reference for all decoration libraries. It's available online at `https://tikz.dev/library-decorations`.

At `https://tikz.net/tag/decorations` and `https://texample.net/tikz/examples/feature/decorations`, you can learn how decorations are beneficial for drawings in science, such as mathematics, mechanics, optics, and electrical engineering.

The Koch snowflake is explained at `https://en.wikipedia.org/wiki/Koch_snowflake`.

9

Using Layers, Overlays, and Transparency

Every TikZ picture is a sequence of drawing commands. As with LaTeX documents, TikZ drawings work linearly. Naturally, we can draw over existing graphic elements; they would overlap, and the most recent drawing action would come on top. We may want something other than overprinting, so we will look at a solution in this chapter.

Until now, we have been drawing on a single canvas. Did you know we can have multiple canvases stacked on each other, such as a background canvas, a regular canvas, and a foreground canvas? These are called **layers** of an image.

In this chapter, you will learn how to utilize layers. Furthermore, you will learn to use see-through paths, such as transparent areas, so that you can see what is behind them.

These are our main topics:

- Using transparency
- Drawing on background and foreground layers
- Overlaying LaTeX content with TikZ drawings
- Positioning pictures on the background of a page

By the end of the chapter, you will be able to draw elements that are not overprinted by others and draw lines or fillings behind already existing nodes or shapes.

Technical requirements

The source code of the chapter's examples is available at `https://tikz.org/chapter-09`.

The code can be downloaded from GitHub at `https://github.com/PacktPublishing/LaTeX-graphics-with-TikZ/tree/main/09-using-layers`.

In this chapter, we will use the `backgrounds`, `matrix`, `quotes`, `positioning`, `fit`, and `decorations.pathmorphing` libraries, and the `tikzducks`, `tikzmark`, `amsmath`, `blindtext`, and `atbegshi` packages.

Using transparency

When you draw anything new that overlaps with previously drawn objects, it simply paints over it so that you cannot see what's behind it anymore. The PDF standard supports partially transparent colors that allow seeing what's behind them.

TikZ provides a simple interface: you can decide how transparent or opaque the colors of an object or path will be by specifying an `opacity` value between 0 and 1. Here, 0 means utterly opaque without transparency, and 1 means entirely transparent, like invisible.

A code is worth a thousand words, so let's have an example. We will draw water, which is naturally transparent. And we'll draw a duck, which is naturally in the water.

To have some waves in the water, we load the `decorations.pathmorphing` library that we used in the previous chapter:

```
\usetikzlibrary{decorations.pathmorphing}
```

TikZ has libraries and packages for everything useful and much silly stuff, so of course, there's a package for drawing ducks. We will play with it in *Chapter 15, Having Fun with TikZ*. For now, let's load it:

```
\usepackage{tikzducks}
```

At the beginning of our `tikzpicture` environment, we draw a smiling duck:

```
\duck[laughing]
```

Then, we draw the water. As planned, it will be transparent; we choose an opacity of 0.5, a blue color gradient, a fill, and a `snake` decoration for getting waves:

```
\fill[top color=white, bottom color=blue, opacity=0.5]
    (-1,-0.2) -- (-1,0.5)
    decorate[decoration={snake}]{-- (3,0.5)} -- (3,-0.2);
```

The two commands give us the following output:

Figure 9.1 – A duck swimming in water

Even though we painted the water over the duck, the duck's body is clearly visible in the water, and that's the transparency we wanted.

We can set the `opacity` value in many ways, such as an option to a node, a path, a scope, or the whole `tikzpicture` environment. In addition, we can choose different opacity settings for lines, fillings, and text, as follows:

- `draw opacity` is for lines and curves.
- `fill opacity` is for filling an area. It is also applied to images and text.
- `text opacity` is for text and especially useful if you want to override the `fill opacity` effect.

Let's create an example to see the effects of different settings.

We will draw a **Venn diagram** to display the intersections of sets. Transparency will help us to visualize the intersections—that is, the overlapping areas. We start our picture with an `opacity` value of 0.4, which means that all that we draw will be 60% transparent:

```
\begin{tikzpicture}[very thick, opacity=0.4]
```

We chose the `very thick` option to see better what happens with lines and curves. Now, we draw three circles of radius 2 with different colors:

```
\filldraw[red]   ( 90:1.2) circle (2);
\filldraw[green] (210:1.2) circle (2);
\filldraw[blue]  (330:1.2) circle (2);
```

The `\filldraw` command draws the circle border and fills it. We used polar coordinates for easy circular arrangement around the origin, at positions 90, 210, and 330 degrees, with a distance of 1.2 to the origin.

We complete the drawing by adding four text nodes, using polar coordinates again:

```
\node at ( 90:2)      {Designing};
\node at (210:2)      {Drawing};
\node at (330:2)      {Coding};
\node [font=\LARGE]   {TikZ};
```

Then, we end the picture:

```
\end{tikzpicture}
```

Compile, and you will get this diagram where everything is transparent:

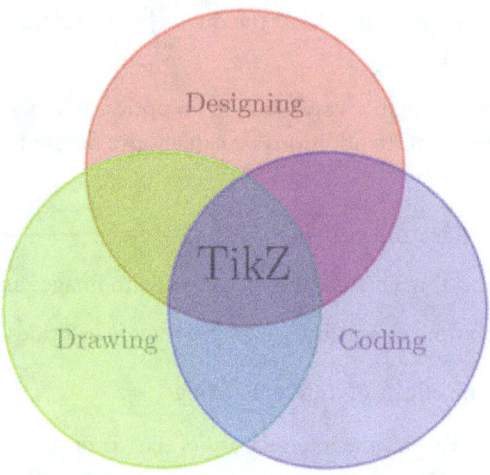

Figure 9.2 – A fully transparent diagram

We can see the following:

- The colors become much lighter. That's not the original red, green, and blue.
- The opacity accumulates; the overlapping areas get darker.
- The text is not black anymore; it's lighter.
- The border of the circles is not solid.

Let's zoom in to investigate the last point:

Figure 9.3 – Borders of areas with transparency

The border segments of the filled circles are two-colored. This is because the filling is drawn precisely to the border and partially overlaps the thickly drawn border. Where filling and border overlap, the accumulating effect darkens the color.

As it looks a bit off, we can fix it in one of the following two ways:

- Add `draw opacity=1` to the `tikzpicture` options so that lines and curves are entirely opaque.
- Change `opacity=0.4` to `fill opacity=0.4` so that lines and curves are unaffected; only filled areas and the text within them get affected.

To ensure that the texts in the nodes are black and opaque, we declare it in the options:

```
\begin{tikzpicture}[very thick, fill opacity=0.4,
  text opacity=1]
```

Both changes result in the following diagram:

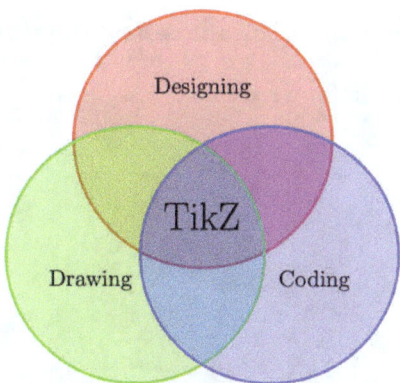

Figure 9.4 – Transparent areas with opaque curves and text

We may use transparent areas, but they should not pile up the effect when they overlap. Look at the following situation: we have a drawing of a "top secret" stamp that consists of a circle and a rectangle; both are transparent:

```
\begin{scope}[opacity=0.6]
   \draw [line width=4mm, red] circle(1);
   \fill[rounded corners, fill=red, rotate=15]
      (-1.3,-0.2) rectangle (1.3,0.2);
\end{scope}
\node[rotate=15] {TOP SECRET};
```

This gives us the following output:

Figure 9.5 – Overlapping transparent objects

We notice that the overlapping areas are of stronger red color. This looks odd; we will have the whole read area with a uniform color. The solution is to use a **transparency group**. We can add this specification as a scope option. Our scope starts now this way:

```
\begin{scope}[opacity=0.6, transparency group]
```

The color in the image changes to a uniform red color:

Figure 9.6 – Grouping transparency

When several objects share the same opacity settings but will not add up the effect when they are overlapping, put them into a `scope` environment and add the `transparency group` option. You may create several scopes for such a purpose in a drawing.

A transparency group works like this: all objects in the scope are drawn one after the other, and if they overlap, then the last element's color is on top. When the scope ends, the whole scope content is drawn altogether with the transparency settings.

We will now proceed to apply our new knowledge to mathematical drawings. We will draw an illustration of creating the so-called **transpose** of a matrix, which is the matrix mirrored on its main diagonal. In this drawing, we will highlight submatrices. At first, we will do it using transparency; later, we will draw on the background layer behind the matrix. Let's set it up.

We load the following TikZ libraries:

```
\usetikzlibrary{matrix,positioning,quotes}
```

We create a standard style for matrices as follows:

```
\tikzset{standard/.style = {matrix of nodes, inner sep=0pt,
  nodes = {inner sep=0.3em},
  left delimiter={(}, right delimiter={)}}}
```

We can define styles in the document preamble; that's recommendable when we have several TikZ pictures and want to use a style several times.

As I love sans serif fonts in diagrams for a clean look without frills, I usually choose the font this way:

```
\tikzset{every node/.append style = {font=\sffamily}}
```

We create two matrices—as we learned in *Chapter 6, Drawing Trees and Graphs*—within a `tikzpicture` environment, of course:

```
\matrix[standard]  (m)   {
    1 & 2 & 3 \\
    4 & 5 & 6 \\
    7 & 8 & 9 \\};
\matrix[standard, right = 3cm of m] (n) {
    1 & 4 & 7 \\
    2 & 5 & 8 \\
    3 & 6 & 9 \\};
```

Finally, we draw an arrow from one matrix to the other:

```
\draw[->,shorten <=1em, shorten >=1em, thick]
   (m.east) to["Transpose"] (n);
```

The previous code altogether gives us this diagram:

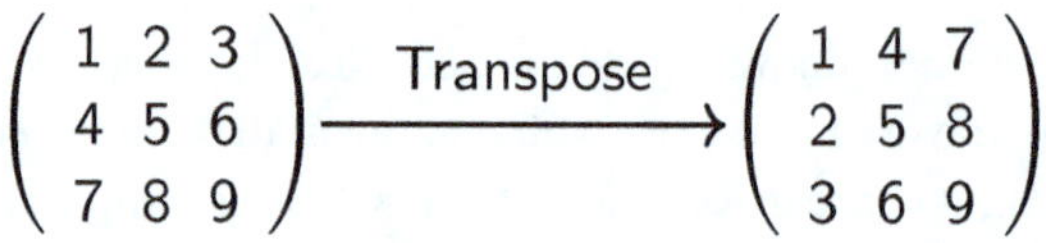

$$\begin{pmatrix} 1 & 2 & 3 \\ 4 & 5 & 6 \\ 7 & 8 & 9 \end{pmatrix} \xrightarrow{\text{Transpose}} \begin{pmatrix} 1 & 4 & 7 \\ 2 & 5 & 8 \\ 3 & 6 & 9 \end{pmatrix}$$

Figure 9.7 – Drawing to display transposing a matrix

On the left hand, we have an example matrix. On the right hand, we see the transpose.

Now, we come to highlighting submatrices. We define a style for this that can also be in the preamble, like the standard matrix style. We choose a yellow color and 50% transparency because otherwise, it would overprint matrix cells:

```
\tikzset{submatrix/.style = {rectangle, rounded corners,
   fill=yellow, fill opacity=0.5}}
```

Now, we draw a rectangle in the submatrix style on the left and again on the right. We use matrix nodes as reference coordinates after we have drawn the preceding m and n matrices in the following commands:

```
\draw[submatrix] (m-2-2.north west)
   rectangle (m-3-3.south east);
\draw[submatrix] (n-2-2.north west)
   rectangle (n-3-3.south east);
```

Figure 9.7 now changes to this:

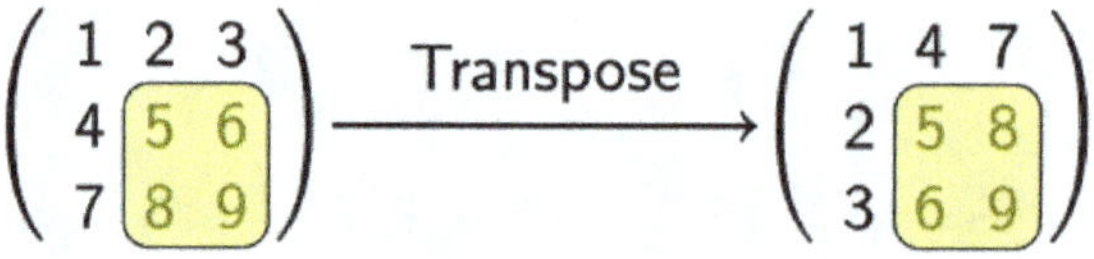

$$\begin{pmatrix} 1 & 2 & 3 \\ 4 & 5 & 6 \\ 7 & 8 & 9 \end{pmatrix} \xrightarrow{\text{Transpose}} \begin{pmatrix} 1 & 4 & 7 \\ 2 & 5 & 8 \\ 3 & 6 & 9 \end{pmatrix}$$

Figure 9.8 – Highlighting a part of a matrix

So, we highlighted submatrices to visualize their change in the matrix transposition.

It would be even better if we had nodes for the submatrices so that we could also draw an arrow between them. TikZ provides the `fit` library for creating and fitting a node to certain coordinates. Let's use that:

```
\usetikzlibrary{fit}
```

We change our `submatrix` style from before so that we can apply it to a node:

```
\tikzset{submatrix/.style = {rectangle, rounded corners,
    fill=yellow, fill opacity=0.5, draw, inner sep=0pt}}
```

Instead of the `\draw` commands for *Figure 9.8*, we now create nodes instead:

```
\node (m1) [submatrix, fit=(m-2-2) (m-3-3)] {};
\node (n1) [submatrix, fit=(n-2-2) (n-3-3)] {};
```

Now, we have nodes with anchors, and we can also draw an arrow between them:

```
\draw [->] (m1.south east)
  to[bend right=20] (n1.south west);
```

You can see that arrow in *Figure 9.9* in the next section, where we continue this example. In the following section, we will draw in the background of the matrix instead of using transparency.

Drawing on background and foreground layers

When a drawing command overprints another object, and we don't want this, we can usually arrange the order of commands. However, it's not always possible. Consider our matrix example from the previous section: We had to draw the highlighting for the submatrices after the matrices because we used their cell coordinates as reference coordinates. We cannot change the drawing order here. Without transparency, the numbers in the cells would be overprinted by the yellow rectangle.

Now, we get to another solution for the overprinting problem: we use **layers**. Specifically, we use the **background layer**. We will apply it to our matrix example.

First, load the `backgrounds` library in the document preamble:

```
\usetikzlibrary{backgrounds}
```

Now, we put the nodes, which we created in the previous section, into a `scope` environment with the `on background layer` option:

```
\begin{scope}[on background layer]
   \node (m1) [submatrix, fit=(m-2-2) (m-3-3)] {};
   \node (n1) [submatrix, fit=(n-2-2) (n-3-3)] {};
\end{scope}
```

We can also remove the `opacity` option in the `submatrix` style. The changed code results in the following output:

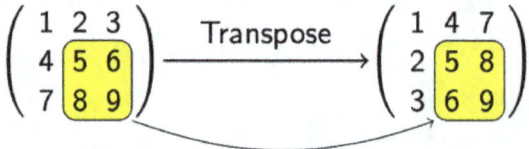

Figure 9.9 – Background highlighting a part of a matrix

You can see that transparency is not used anymore; we have the full yellow color, and it doesn't overprint the matrix cell content, even though the node commands come later than the matrix in the code.

Compare *Figure 9.9* with *Figure 9.8*: another significant change is that the color of the text in the matrix is not affected by the highlighting and remains solid black.

While the background layer is the most interesting in drawing, you can use several layers. You don't need to load a TikZ library, as the `pgf` backend provides the features we need.

At first, you declare the layers you want to use, such as the following:

```
\pgfdeclarelayer{background}
\pgfdeclarelayer{foreground}
```

The default layer is called `main` and is available by default.

Then, you can define the order in which they are stacked on top of each other:

```
\pgfsetlayers{background,main,foreground}
```

Now, you can draw on each layer in the following way, with the `foreground` layer as an example:

```
\begin{pgfonlayer}{foreground}
   \node {Some text};
\end{pgfonlayer}
```

In the next section, we go one step further: we will add TikZ drawings to standard LaTeX text and math content.

Overlaying LaTeX content with TikZ drawings

Our matrix example was fine for creating a diagram. However, mathematicians will typeset matrices using `amsmath` matrix environments in LaTeX's math mode.

In this section, we will explore how to draw within such text and math content created outside TikZ. There will be quite some code lines again; remember that you can view the complete example code at `TikZ.org` and on GitHub.

We will recreate our matrix example with standard LaTeX tools. First, we have to load the `amsmath` package. Then, in the document, we write an equation with `pmatrix` environments and a long extensible arrow. That's straightforward LaTeXing. One thing will be new here: whenever we want to remember a current position, we insert the `\tikzmark{x}` command, where x stands for a coordinate name we choose. Let's go:

```
\[
  \begin{pmatrix}
     1 & 2 & 3 \\
     4 & \tikzmark{m1}5 & 6 \\
     7 & 8 & 9\tikzmark{m2}
  \end{pmatrix}
  \xrightarrow{\text{Transpose}}
  \begin{pmatrix}
     1 & 4 & 7 \\
     2 & \tikzmark{n1}5 & 8 \\
     3 & 6 & 9\tikzmark{n2}
  \end{pmatrix}
\]
```

The new command comes from a TikZ library with the same name. To use this and TikZ in our math document, we load TikZ and the required libraries and define our `submatrix` style:

```
\usepackage{tikz}
\usetikzlibrary{fit,tikzmark}
\tikzset{submatrix/.style = {draw,rectangle,
  rounded corners, fill=yellow}, inner sep=2pt}
```

Now comes the fun part. The `tikzmark` package provides the `pic` coordinate system; we can refer to our new coordinates by `(pic cs:m1)`, `(pic cs:m2)`, `(pic cs:n1)`, and `(pic cs:n2)`. We can create a TikZ drawing using those coordinates as long as we are on the same page. So, we place it either before or after the equation—depending on if the drawing will be in the background of the equation—or overprint it. *Important*—we have to use two options:

- `overlay` so that our TikZ picture doesn't require space for it but is overlaying
- `remember picture` so that TikZ remembers picture positions from the previous compiler run

So, put this code right before the equation with the matrices:

```
\begin{tikzpicture}[overlay, remember picture]
  \node (m) [submatrix,
    fit={([yshift={1.5ex}]pic cs:m1) (pic cs:m2)}] {};
  \node (n) [submatrix,
    fit={([yshift={1.5ex}]pic cs:n1) (pic cs:n2)}] {};
  \draw [->] (m.south east) to[bend right=20] (n.south west);
\end{tikzpicture}
```

Since `\tikzmark` is unaware of the text around it, the coordinate is at the baseline. We inserted `[yshift={1.5ex}]` to shift the coordinate a bit upward. That's a way to tweak it, and it's a small price for overlaying arbitrary content with TikZ drawings.

We have to compile twice because LaTeX and TikZ have to gather the coordinates in the first compiler run before they can be used. Then, we get this output:

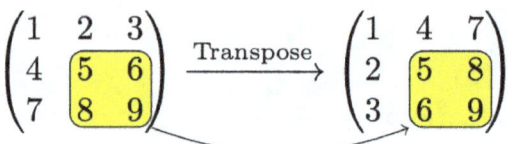

Figure 9.10 – Drawing in the background of a LaTeX equation

The main benefit of this method is that we can add TikZ content to any LaTeX content that's written using regular LaTeX or any other LaTeX package.

In the next section, we will explore the `remember picture` option further.

Positioning pictures on the background of a page

A regular TikZ picture appears right where we put it in the document text. We can use a `center` environment for centering it and a `figure` environment to have a caption and a reference label and to let it float to a suitable place in the document for better page breaks.

We can even put a TikZ picture anywhere on a page without affecting the document text, placing it in the background. As we already know, we can use the `overlay` option so that it doesn't block any space. The `remember picture` option from the previous section is even more important here: once you set this, TikZ stores picture positioning information in the `.aux` file. In the next compiler run, that information is read from the `.aux` file and used for final positioning. That's the reason why we need to compile a second time to have the final positioning.

We can consider the page like a node with a rectangular shape, having precisely the page dimensions. TikZ calls it the `current page` node, and we can use the node anchors in the same way we did in *Chapter 3*, *Drawing and Positioning Nodes*. For example, (`current page.center`) is the coordinate for the middle of the page, while (`current page.north east`) is the top-right corner. We can use those coordinates as reference points for absolute or relative positioning, as we did in *Chapter 3*.

Let's have a simple but complete example to see it in action. We will create a small LaTeX document with dummy text. We aim to place a very big gray **DRAFT** sign across the page.

So, our TikZ picture is just a single node with text, gray color, enlarged by some `scale` factor, and rotated. That's a good occasion to practice the short `\tikz` command instead of the `tikzpicture` environment.

The following code will give us a large, rotated **DRAFT** sign at the page center:

```
\tikz[overlay,remember picture]
  \node[rotate=40,scale=10,lightgray,font=\bfseries]
    at (current page.center) {DRAFT};
```

If we use this code before we write the page text content, it's placed in the background behind the text. If we put this code after we wrote the text on the page, it would overprint the text. In the latter case, if we prefer it, we may add an opacity value to the node options to get it transparent.

Now, let's see how we can place the **DRAFT** sign on the background of every document page. We can add it to the process when TeX finishes generating the page content and does the actual output. This process is called **shipout**. We will use a package that allows us to add commands *at the beginning of the shipout* of a page. The package name is an abbreviation for this task; it's called `atbegshi`.

We will use two commands from that package:

- `\AtBeginShipout{code}`: This tells LaTeX to execute code for every page.
- `\AtBeginShipoutAddToBox{code}`: This adds code to the page box without dimensions, as we had with `overlay`. This is where we place our TikZ code.

For more details and further commands, take a look at the package documentation. We will load the package and use both commands, as highlighted in the following example:

```
\documentclass{article}
\usepackage[english]{babel}
\usepackage{tikz}
\usepackage{blindtext}
\usepackage{atbegshi}
\AtBeginShipout{\AtBeginShipoutAddToBox{%
  \tikz[overlay,remember picture]
    \node[rotate=40,scale=10,lightgray,font=\bfseries]
      at (current page.center) {DRAFT};}}
\begin{document}
\blinddocument
\end{document}
```

The `\blinddocument` command from the `blindtext` package generates a dummy document containing some sections with text. As you know, we need to compile the document twice. Then, the first two pages look like the following, with the addition of our **DRAFT** sign:

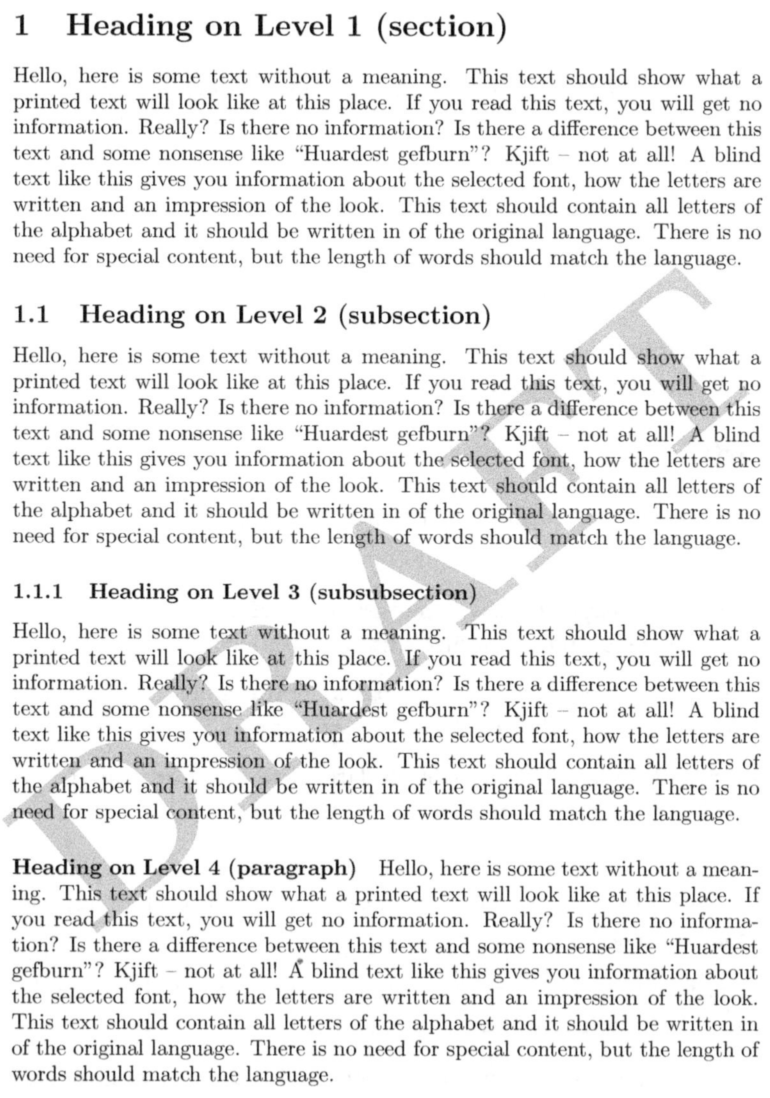

1 Heading on Level 1 (section)

Hello, here is some text without a meaning. This text should show what a printed text will look like at this place. If you read this text, you will get no information. Really? Is there no information? Is there a difference between this text and some nonsense like "Huardest gefburn"? Kjift – not at all! A blind text like this gives you information about the selected font, how the letters are written and an impression of the look. This text should contain all letters of the alphabet and it should be written in of the original language. There is no need for special content, but the length of words should match the language.

1.1 Heading on Level 2 (subsection)

Hello, here is some text without a meaning. This text should show what a printed text will look like at this place. If you read this text, you will get no information. Really? Is there no information? Is there a difference between this text and some nonsense like "Huardest gefburn"? Kjift – not at all! A blind text like this gives you information about the selected font, how the letters are written and an impression of the look. This text should contain all letters of the alphabet and it should be written in of the original language. There is no need for special content, but the length of words should match the language.

1.1.1 Heading on Level 3 (subsubsection)

Hello, here is some text without a meaning. This text should show what a printed text will look like at this place. If you read this text, you will get no information. Really? Is there no information? Is there a difference between this text and some nonsense like "Huardest gefburn"? Kjift – not at all! A blind text like this gives you information about the selected font, how the letters are written and an impression of the look. This text should contain all letters of the alphabet and it should be written in of the original language. There is no need for special content, but the length of words should match the language.

Heading on Level 4 (paragraph) Hello, here is some text without a meaning. This text should show what a printed text will look like at this place. If you read this text, you will get no information. Really? Is there no information? Is there a difference between this text and some nonsense like "Huardest gefburn"? Kjift – not at all! A blind text like this gives you information about the selected font, how the letters are written and an impression of the look. This text should contain all letters of the alphabet and it should be written in of the original language. There is no need for special content, but the length of words should match the language.

Figure 9.11 – Drawing in the background of LaTeX pages

As the picture is drawn before the document text on the page, we don't need transparency here.

In the same way, you can draw to the page foreground instead, thus overprinting the page content. Just use `\AtBeginShipoutAddToBoxForeground` instead of `\AtBeginShipoutAddToBox`. They are quite long command names, and I thought TikZ would be verbose. In this case, using transparency is recommendable.

Summary

Having worked through this chapter, you now master overlapping situations and control the order of drawing commands in your source code versus the order of appearance in the output image.

Now, you can draw see-through content using transparent colors and draw on the background of pictures behind other objects without overprinting them.

You learned to utilize the `tikzmark` library, which is a very clever and helpful tool for tweaking your LaTeX articles, books, or presentation slides with TikZ drawing tools.

You can now draw on absolute positions on a page behind the regular LaTeX document text. In the next chapter, you will learn to calculate with coordinates, which helps a lot in positioning.

Further reading

The following sections in the TikZ manual at `https://texdoc.org/pkg/tikz` are relevant to this chapter:

- *Part III, Section 23.1, Transparency* covers transparency, blending colors, and so-called **fadings**—non-uniform transparency. The online link is `https://tikz.dev/tikz-transparency`.
- *Part V, Section 45, Background Library* is the `backgrounds` library reference and is available online at `https://tikz.dev/library-backgrounds`.
- *Part IX, Section 113, Using Layers* explains declaring and using layers in the basic PGF layer, to be found online at `https://tikz.dev/base-layers`.

The TikZ galleries (`https://tikz.net` and `https://texample.net/tikz/examples`) contain many examples using features of this chapter; those collections have **layers**, **background**, **fadings**, and **transparency** tags.

The `tikzmark` manual can be found at `https://texdoc.org/pkg/tikzmark`. The `atbegshi` manual is at `https://texdoc.org/pkg/atbegshi`.

The transpose of a matrix, for which we made drawings in this chapter, is explained on *Wikipedia* at `https://wikipedia.org/wiki/Transpose`.

10
Calculating with Coordinates and Paths

In *Chapter 1*, *Creating the First TikZ Images*, we started off using explicit values to choose coordinates. We achieved relative positioning by manually giving distances while drawing a path.

Now, we are about to take things to the next level by introducing a whole new set of techniques for calculating coordinates from other coordinates. We can add or subtract coordinates from each other, calculate a coordinate position between other coordinates at a certain distance, find a coordinate as a projection onto a line, and rotate coordinates.

And that's not all – we'll introduce loop commands that help repeat calculations and actions.

Get ready to dive deep into the following techniques:

- Repeating in loops
- Calculating with coordinates
- Evaluating loop variables
- Calculating intersections of paths

By the end of this chapter, you will be ultra-efficient in repeating similar commands and using calculations for perfect node and edge placement.

Technical requirements

The source code of the chapter's examples is available at https://tikz.org/chapter-10. The code can be downloaded from GitHub at https://github.com/PacktPublishing/LaTeX-graphics-with-TikZ/tree/main/10-calculating-transforming.

In this chapter, we will use the `calc` and `intersections` TikZ libraries; other features are loaded by default.

Repeating in loops

The easiest calculation is counting, so this will be our starting point. In a **for loop**, TikZ can count with a **variable** for us while it repeats a code segment using the variable. While this sounds simple, it's tremendously valuable for generating graphics with ease, especially with the TikZ \foreach command, which is incredibly flexible.

The basic syntax of this command is the following:

```
\foreach variable in {list of values} {commands};
```

Let's break down the highlighted code:

- `variable`: We name and use it like a macro, such as \i. The convention of using *i* as a loop variable dates back to the early programming languages and mathematics, when *x* and *y* were used for variables and *i* and *j* were used as indexing counters. However, we are free to choose any name as long as it starts with a backslash.
- `list of values`: This is a comma-separated list of values, such as 1,2,3. You can omit values and write – for example, 1,...,10 – and then TikZ implicitly fills in the missing values – here, all numbers from 1 to 10. When you give more values, TikZ calculates the difference and uses it for filling in. So, with 2,4,...,10, TikZ uses the even numbers until 10. That auto-filling works even with fractional steps, such as 0.1,0.2,...,1. Plus, you can use alphabetic character sequences and patterns such as A_1,...,F_1.
- `commands`: This can be a sequence of commands that use the variable. If you use a single command, you can skip the braces around it.

Let's look at the syntax with some real examples. When we take the grid from *Figure 2.1*, we can add labels to the *x* axis like this:

```
\foreach \i in {-3,-2,-1,1,2,3} \node at (\i,-0.2) {\i};
```

As we can use several commands in a single loop, we can add *x* and *y* labels at the same time:

```
\foreach \i in {-3,-2,-1,1,2,3} {
  \node at (\i,-0.2) {\i};
  \node at (-0.2,\i) {\i};
}
```

Now our grid looks like this:

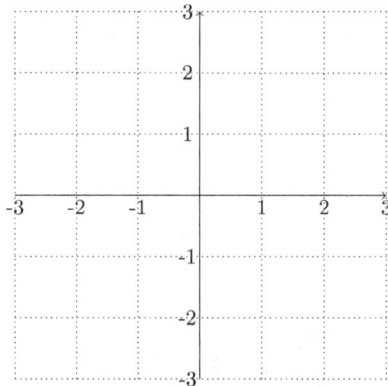

Figure 10.1 – A grid with axis labels

That's useful when we use this draft helper grid in more extensive drawings with a larger grid.

By using the dots auto-filling feature, we can write it shorter, such as the following:

```
\foreach \i in {-3,...,3} \node at (\i,-0.2) {\i};
```

This would, of course, also include the value 0.

Let's also see how the auto-filling of omitted values works. We want to draw 36 circles at a distance of 1 to the origin, every 10 degrees between 10 and 360 degrees. It's sufficient to tell TikZ to start the loop with 10, proceed with 20, and continue that way until 360 is reached:

```
\foreach \i in {10,20,...,360} \draw (\i:1) circle (1);
```

That single line gives us a set of circles with nice symmetry:

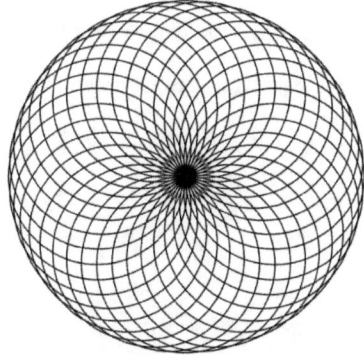

Figure 10.2 – Rotated circles

In the code line for *Figure 10.2*, we got 36 separate drawing paths. Let's say we want to fill all these circles, not simply black but alternating black and white. Remember the `even odd` filling rule from *Chapter 7*? We can apply that filling when we turn this example into a single path. Luckily, we can use `\foreach` within a single path. We can change our example as follows:

```
\filldraw[even odd rule] \foreach \i in {10,20,...,360}
   {(\i:1) circle (1)};
```

With the `even odd` filling rule, adjacent areas of a self-intersecting path have different colors, and so we get this amazing pattern as a result:

Figure 10.3 – Filled intersecting circles

You can see that with a single `\foreach` line, you can draw a lot with many iterations.

`\foreach` can take several loop variables and values, separated by forward slashes, as follows:

```
\foreach \i/\j in {A/1,B/2,C/3} \node at (\j,-0.2) {\i};
```

This prints **A**, **B**, and **C** instead of 1, 2, and 3 at the *x* axis shown in *Figure 10.1*, as follows:

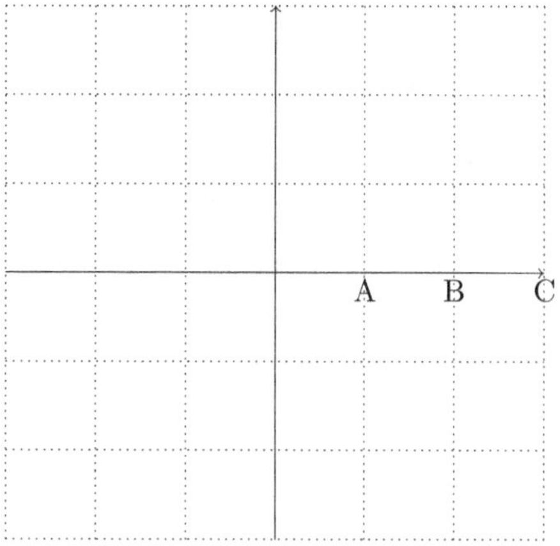

Figure 10.4 – Alphanumeric labels

Whenever you have a command that you would like to execute repeatedly for a specific set of values, pairs, or triples, you can use the \foreach command. And here's the best part – it's more than just handy in TikZ; you can use \foreach directly within LaTeX. To use it without loading TikZ, all you need to do is to load the pgffor package where it's defined.

Now that we've covered the basics of the \foreach command, let's quickly advance to the next topic because, in the following sections, we can see further examples. Loops are even more powerful when we combine them with calculations, and that's what we'll be exploring next.

Calculating with coordinates

The calc library allows us basic operations with coordinates. Load it first in the document preamble with \usetikzlibrary{calc}, and you're ready to go.

TikZ can now calculate what we write between $ symbols within a coordinate. While it may look like TeX inline math mode, it actually enables us to perform calculations and math operations.

Adding and subtracting coordinates

With just the simple notation of $ (A) + (B) $ we can add two coordinates. How can this be useful? It's an easy relative positioning when we use a particular coordinate and add a coordinate to have a shift in the *x* and *y* directions.

Let's start with a coordinate, A, at some arbitrary position and build what we can see in *Figure 10.5*:

```
\coordinate (A) at (1,2);
```

Now, we can create a coordinate that's just right of it, with an x distance of 1, by adding a coordinate with suitable values, $x=1$ and $y=0$:

```
\coordinate (B) at ($(A)+(1,0)$);
```

Note that we also have parentheses around the $...$ expression to indicate that it's a coordinate. Admittedly, the nesting of parentheses and $ symbols looks messy, but it's well structured.

We can also use polar coordinates. The following line creates a coordinate, C, which also has a distance of 1 to A, but with an angle of 60 degrees to the x axis, by adding a coordinate in polar notation:

```
\coordinate (C) at ($(A)+(60:1)$);
```

When we connect A, B, and C, we get an equilateral triangle:

```
\draw (A) -- (B) -- (C) -- cycle;
```

In the same way, we can subtract coordinates from each other. Furthermore, we can insert a factor expression before coordinates with a * symbol, which can be a number or even a more or less complex computation, such as in 2*(A), sqrt(3)*(2,2). Such a calculation can be performed like this:

```
\coordinate (D) at ($sin(60)*sqrt(2)*(A)+0.5*(60:1)$);
```

You may rarely need it, but it's good to know that math tools are available if you need them.

Computing points between coordinates

We can save effort and let TikZ calculate the position of points between two coordinates. The basic syntax is like (A)!factor!(B), which gives a coordinate on the line between A and B, with the factor between 0 and 1 deciding where – (A)!0.1!(B) is close to A, (A)!0.9!(B) is close to B, and (A)!0.5!(B) is precisely the midpoint between A and B. A factor of 0 would simply equate to A, and a factor of 1 would be equivalent to B.

We are allowed to use negative values and higher values than 1; in that case, the resulting point will still be on the line between A and B, but not next to them. When we use negative factors, the new coordinate will have the same distance from A as the corresponding positive factor but lie in the other direction on the line, away from both A and B. So, (A)!2!(B) would be twice as far from A as B is.

TikZ calls this kind of expression a **partway modifier**.

To see the syntax in use, we can draw the inscribed circle of the ABC triangle from the previous section in *Figure 10.5*. A bit of math research reveals that the radius shall equal sqrt(3)/6; we use that fact to draw the circle above the middle point between A and B:

```
\draw ($(A)!0.5!(B)+(0,{sqrt(3)/6})$) circle ({sqrt(3)/6});
```

Note how we used curly braces to encapsulate the math expression. Generally, adding braces helps us when the parser gets confused by additional syntax because, as you can see, we can have pretty complex math expressions. Here, it was particularly needed because the parser expects parentheses for coordinates and would get confused by the parentheses of the `sqrt` function.

This is the result of our drawing with calculated coordinates in this section:

Figure 10.5 – A triangle with an inscribed circle

Instead of a factor, we can use a dimension; otherwise, the syntax stays the same. So, `(A)!1cm!(B)` is the point on the line between A and B with a distance of 1 cm. Similarly, `(A)!-1cm!(B)` is the point on the line connecting A and B, which is not between them but on the other side of A, with a distance of 1 cm. That's straightforward and called a **distance modifier**.

Projecting on a line

The third expression with very similar syntax is the **projection modifier**. Instead of a factor or a distance, we can insert a third coordinate. Let's say the third coordinate is C; then, `(A)!(C)!(B)` is the orthogonal projection from C onto the line connecting A and B. It doesn't have to be between A and B.

Here, you can see it in action with the previous example, drawing a dotted line from C to the orthogonal projection from C on the line between A and B:

```
\draw[densely dotted] (C) -- ($(A)!(C)!(B)$);
```

Figure 10.5 with the additional line now looks like this:

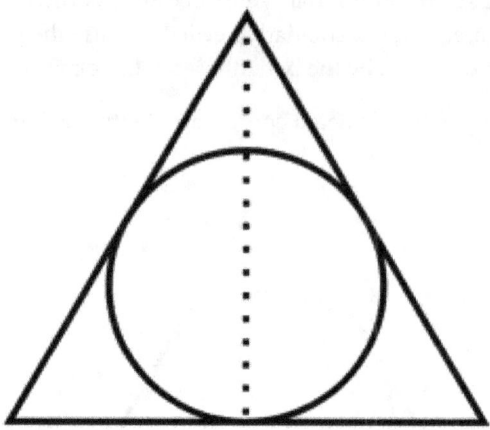

Figure 10.6 – The projection on a line

Of course, you can combine a projection with further calculations, such as factors and angles, which we will see in the next section.

Adding angles

With all modifiers, we can insert an angle. That's a value in degrees prefixed to the second coordinate, separated by a colon. At first, the line from A and B would be rotated by that angle around A, and then the modifier would be applied.

So, with our example, the full expression, ($(A)!0.5!60:(B)$), equals the coordinate right in the middle between A and B, rotated by 60 degrees around A.

We can apply it to our equilateral triangle in the following way:

```
\filldraw ($(A)!0.5!60:(B)$) circle (0.03);
```

As each angle of the triangle happens to be 60 degrees, it's equal to the middle point between A and C, as we can see here:

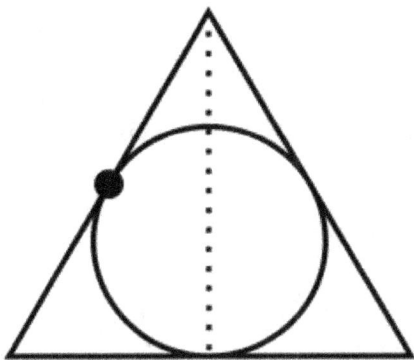

Figure 10.7 – Using a partway modifier with an angle

Let's practice the partway modifier with angles and a \foreach loop to get a glimpse of practical code. We will arrange circles in an Archimedean spiral. That is a spiral of polar coordinates where the radius is proportional to the angle. We start with an angle of zero, having a radius of zero. At half of the full 360 degrees, we have a radius of 0.5. At 360 degrees, we have a radius of 1. That can continue, so with 720 degrees, we will have a radius of 2, and so on.

We will make a \foreach loop with tiny steps to get small circles; we will iterate over a \i variable that shall be the fraction of the angle. \i will be our partway modifier between the origin, (0,0), and the coordinate, (1,0). \i will also serve as the fraction of 360 degrees. The circle radius shall also grow with \i; we will add a suitable factor so that it's small enough. That's the plan, and now here's the code; try to understand it with the preceding explanation:

```
\foreach \i in {0,0.025,...,1}
   \draw ($(0,0)!\i!\i*360:(1,0)$) circle(0.08*\i);
```

We get one rotation of the spiral:

Figure 10.8 – A spiral of circles

To see the six spiral rotations, we can let \i run in the loop until 6. Add shading to get colored balls:

```
\foreach \i in {0,0.025,...,6}
  \draw[shading=ball] ($(0,0)!\i!\i*360:(1,0)$)
    circle(0.08*\i);
```

Without much work, we get an impressive output thanks to the loop:

Figure 10.9 – A spiral of balls

In the next section, we will see how to calculate within \foreach options.

Evaluating loop variables

Let's take a closer look at the code for *Figure 10.8*. Could there possibly be even more variable options to adjust? How about iterating colors and interconnecting between loop repetitions? Yes, you can achieve this within the same loop. So, let's look at further advanced \foreach options – at first, some syntax with short examples, and then a more extended example.

Counting loop repetitions

A \foreach value list can contain alphanumeric values and patterns to be iterated through and utilized within the loop commands. However, we may want to use values based on their position in the list, such as using the (\j,0) coordinate at position j in the iterative order of list values.

This is where the count option comes into play. Let's say we have \i as the loop variable iterating through letters. We introduce the \j counter as follows:

```
\foreach \i [count=\j] in {A,...,Z} {commands};
```

Now, while \i iterates from A to Z, \j goes from 1 to 26, and we can use both \i and \j in the commands.

We don't have to start with 1. By writing `count=\j from 10`, we let \j begin at 10 instead of 1.

Of course, we can choose any name instead of \i and \j.

Evaluating the loop variable

As the loop variable can be some pattern, it is not evaluated to a number by default. It is used in the commands as it is, without pre-calculating its value. We can force it or even do a complicated custom computation. It works as follows; again, \i and \j are chosen as names:

```
\foreach \i [evaluate=\i as macro using formula]
   in {values} {commands};
```

`macro` is our additional variable name, such as \j, and `formula` can be a math expression.

If we only say `evaluate=\i`, that value is used when we use \i in the commands. If we only say `evaluate=\i as \j`, then \i stays as its original pattern, and \j is the evaluated value for using both in the commands.

When we use the full syntax with `formula`, this formula will be used for evaluation \j, with some math expression applied to \i. We will practice it at the end of this section.

Remembering the loop variable

When we have any repetition in a loop, we may want to remember the variable value from the previous repetition, such as to connect points. That's done as follows:

```
\foreach \i [remember=\i as macro initially value]
   in {values} {commands};
```

`macro` can be a name that we choose, such as \j. Now, \j will have the value of \i from the previous repetition. The initial `value` is the value at the first repetition when there is no last value.

To practice these evaluations, we will modify the example for *Figure 10.8* as follows:

```
\foreach \i [remember=\i as \j (initially 6),
   evaluate=\i as \c using 20*\i] in {5.95,5.9,...,0}
   \fill[fill=black!60!blue!\c!white]
      ($(0,0)!\i!\i*180:(1,0)$) --
      ($(0,0)!\j!\j*180:(1,0)$) -- (0,0);
```

The following happens:

- \i is our loop variable, this time demonstrating that we can do negative steps. We do tiny steps of 0.05, starting lower than 6 and going to 0.
- \j is the remembered previous value of \i in each loop repetition, starting from 6.
- \c is the color we use for filling the evaluation based on the value of \i; the blue value gets lighter in each repetition.
- The loop commands fill a triangle based on the \i and \j values as corners, with the origin (0,0) as the third corner, in a calculation like that shown in *Figure 10.8*.

We get the following as output:

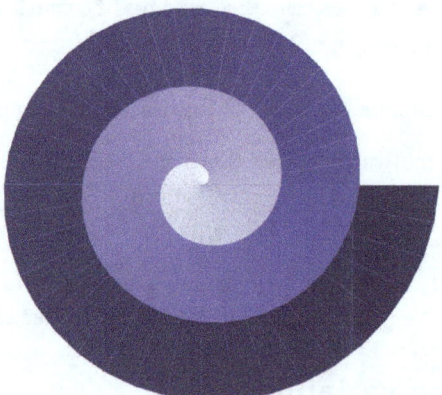

Figure 10.10 – A colored segmented spiral

That was already pretty complex. If you need even more flexibility, remember that you can have several loop variables and nest the \foreach loops. Experiment with it, and share your examples on the TikZ community gallery site: https://tikz.net.

In the next and final section, we will generate coordinates from existing paths.

Calculating intersections of paths

TikZ drawings are often built step by step. We choose coordinates and draw lines, curves, and shapes. At some point, we may need to know the intersection of such paths to proceed with further drawing steps, such as adding text or arrows at such positions.

We could calculate the intersection point of two lines ourselves by solving a system of two linear equations. To get the intersection points of a circle and a line, we can solve a quadratic equation. Remember polygons or shapes consisting of curvy paths such as bent lines? It can become hard to compute a point on such a path that overlaps with another path.

TikZ provides the intersections library that solves such challenges. You can load it in the usual way:

```
\usetikzlibrary{intersections}
```

Now, TikZ can do all the hard work and calculate all intersection points of arbitrary paths, generating named coordinates for them.

Let's dive into a basic example to see how it works. We'll need to use **named paths**, which means we will declare the name as an option for each path. The following code draws two lines called l1 and l2:

```
\draw[name path = l1]  (-2,-2)  --  (3,3);
\draw[name path = l2]  (-1,3)   --  (3,-3);
```

The name intersections option generates the intersection coordinates, which are named intersection, followed by a dash and a number starting from 1. We need to specify the paths using the of keyword, as follows:

```
\fill[name intersections = {of = l1 and l2}]
    (intersection-1) circle(1mm) node[right] {here};
```

That command draws a bullet at the intersection of lines l1 and l2, with text next to it. Together with our helper grid from *Figure 2.1*, it looks like this:

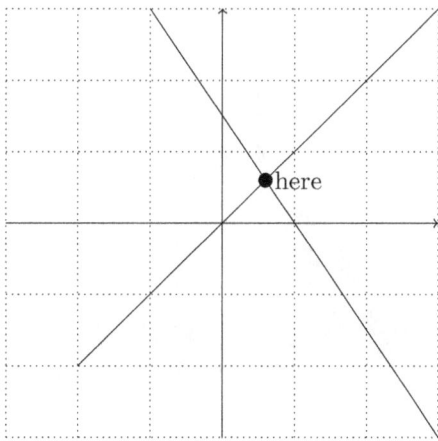

Figure 10.11 – A point at the intersection of two lines

This is the list of available intersection library keys and options:

- name path: This is the name we give to the ongoing path in the current scope. Use name path global if you need the path names beyond scopes.
- name intersections: That's a list of options in curly braces.

- `of`: Here, you specify the names of paths, together with the and keyword.
- `name`: You can select an optional prefix to replace the default intersection prefix.
- `total`: This is a macro name that stores the total number of intersections TikZ found, useful for iterating through in a `foreach` loop.
- `by`: Here, you can write a comma-separated list of coordinates that you want to use for the intersection points, such as a, b, and c, instead of `intersection-1`, `intersection-2`, and `intersection-3` respectively. Like in a `foreach` list, you can use the ... notation.
- `sort by`: You can state the name of the path that shall be the reference for sorting the intersection coordinates, instead of the order in which TikZ found them.

We can create a more sophisticated example with complex paths and more intersection points to see that syntax come alive. In *Figure 7.9*, we had a circle intersecting triangle paths. Let's use this and choose two circles that overlap with the two triangles, as follows:

```
\fill[name path=triangle, orange]
   (90:2) -- (210:2) -- (330:2) -- cycle
   (90:1) -- (330:1) -- (210:1) -- cycle;
\draw[name path=circle, dashed, gray]
   circle(1.5) circle(0.65);
```

This gives us the following:

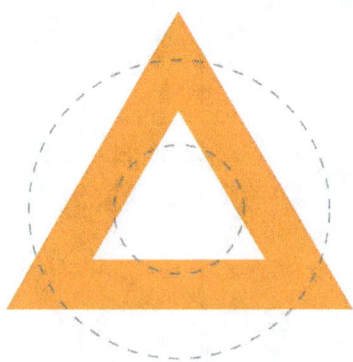

Figure 10.12 – A filled triangle path with intersecting circles

Now, we use the preceding keys and options to find, sort, and number all intersection points:

```
\fill[blue,
   name intersections = {of = triangle and circle,
   total=\max, name=c, sort by = circle}]
```

```
\foreach \i in {1,...,\max} {
  (c-\i) circle(0.5mm)
    node[above left=0.5mm,font=\tiny, inner sep=0]{\i}};
```

Now, TikZ finds \max=12 intersection coordinates, sorted in the order of the circle's path, counterclockwise, and we can use the short c prefix for compact notation. That command adds the points and labels as follows:

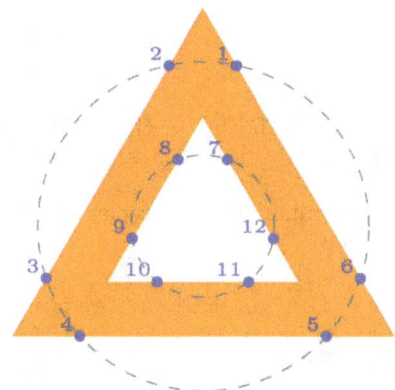

Figure 10.13 – Intersections of circles and triangles

You can see that even though our paths are not continuous but instead consist of multiple shapes, TikZ finds all intersection points with ease, regardless of the path complexity.

Summary

You saw that TikZ's loops are incredibly flexible. You can use loops whenever you see stuff repeating and want to save yourself from writing repetitive code.

Calculating with coordinates can now make your life easier. Just add a coordinate to move in the x or y direction or with an angle or distance. Use factors to place something in between nodes or coordinates. That's not just for geometry; this handy syntax is helpful for any lines, arrows, or positioning nodes in complex diagrams in a perfectly controlled manner.

Letting TikZ calculate intersection points of lines, curves, and complex paths helps you create more intricate shapes based on simpler ones.

In the next chapter, you will learn how to transform coordinates, paths, and scopes, such as by transposition and rotation.

Further reading

The following sections in the TikZ manual at https://texdoc.org/pkg/tikz are the reference for the commands, syntax, and libraries used in this chapter:

- *Part VII, Section 88, Repeating Things: The Foreach Statement*, gives all formal details of the \foreach command and its syntax. You can also find it at https://tikz.dev/pgffor.
- *Part III, Section 13.5, Coordinate Calculations*, is the reference for the calc library. The direct online link is https://tikz.dev/tikz-coordinates#sec-13.5.
- *Part III, Section 13.3, Coordinates at Intersections*, explains working with path intersections and is available online at https://tikz.dev/tikz-coordinates#sec-13.3.

As you already know, the TikZ galleries contain many examples relevant to this chapter. You can visit the following:

- https://texample.net/tikz/examples/feature/foreach
- https://tikz.net/tag/foreach
- https://texample.net/tikz/examples/feature/coordinate-calculations
- https://tikz.net/tag/calc

The Archimedean spiral is explained on Wikipedia at https://wikipedia.org/wiki/Archimedean_spiral.

11
Transforming Coordinates and Canvas

In this chapter, we will deal with **transformations**. We already used a basic transformation in the code for *Figure 9.0* when we used the `yshift` option to move a coordinate higher in the y-direction. We will now look thoroughly into moving, rotating, and scaling coordinates and apply this to our drawings.

In particular, we will explore the following topics:

- Shifting nodes and coordinates
- Rotating, scaling, and slanting
- Transforming the canvas

Once you have mastered this chapter, you will be equipped with a toolset for minor adjustments and even complex relative positioning of coordinates, nodes, edges, or complex paths.

Technical requirements

At `https://tikz.org/chapter-11`, you can study and compile the complete code for this chapter's examples. The GitHub link for downloading the code is `https://github.com/PacktPublishing/LaTeX-graphics-with-TikZ/tree/main/11-transforming-coordinates`.

This chapter doesn't require additional TikZ libraries, though you may take a look at the `tikz-ext` package.

Shifting nodes and coordinates

We will start with **shifting**. This is another word for doing a **translation**, which means moving a coordinate or a node to another position in a straight line by a particular dimension or by another coordinate.

The following three options can be used for shifting:

- `xshift` is a dimension for moving in the x-direction, adding this dimension to the x value of the coordinate.
- `yshift` does the same but in the y-direction. Here, TikZ adds this dimension to the y value of the coordinate.
- `shift` is a coordinate to be added; its x and y values will be added to the x and y values of the other coordinate. Here, no dimension is used. You can use dimensions, but you don't have to. The coordinate must be given in curly braces.

The shifting value can be used as an option to a path, so it's applied to every coordinate in the path. For example, the following command draws a line from (0,2) to (1,3):

```
\draw[yshift=2cm] (0,0) -- (1,1);
```

You can combine the options. The following code line draws a circle of radius 1 with the center at (1,2):

```
\draw[xshift=1cm, yshift=2cm] (0,0) circle(1);
```

Using the `shift` option, that line does the same and is shorter:

```
\draw[shift={(1cm,2cm)}] (0,0) circle(1);
```

Writing `shift={(1,2)}` would do the same if you did not change the unit length's default value of 1 cm. But remember to enclose the coordinate in curly braces so TikZ can parse it correctly. Specifically, it does not mistake the comma in the coordinate with a comma separating options.

You can apply such transformation options to several paths at once using a `scope` environment.

You can apply the shifting to particular coordinates within a path by using the option in square brackets within the coordinate right after the opening parenthesis. The following command draws a line from (3,4) to (4,5):

```
\draw ([shift={(2,3)}]1,1) -- (4,5);
```

In *Chapter 10, Calculating with Coordinates and Paths*, you learned about the `calc` syntax. The following command does the same as the previous one:

```
\draw ($(1,1)+(2,3)$) -- (4,5);
```

You may think all of that looks so complicated. Why not simply use the final coordinate right from the beginning? The reason is that we have many coordinates in drawings where we don't know the value, such as anchors, positions on edges, or intersections and projections. We can transform the value of such implicit coordinates to place nodes or other objects using the options we learn about in this chapter. For example, the following \draw command puts a node in between two nodes, A and B, slightly shifted up so the text doesn't overprint the line:

```
\draw (A) -- (B) node[pos=0.5, yshift=2mm] {text};
```

You see that you can apply a transformation to the position of a node to be placed. However, we cannot simply transform based on existing nodes. The following line has no shifting effect; it just draws a line from A to B:

```
\draw[yshift=2cm] (A) -- (B);
```

However, you can refer to the node's anchor coordinates like this:

```
\draw ([yshift=2cm]A.east) -- ([yshift=2cm]B.west);
```

Note that even if you defined A and B using \coordinate, such as \coordinate (A) at (0,1), they are actually nodes with empty text and not TikZ coordinates, so the same issue applies to them. But of course, here you can also refer to anchors, such as (A.center).

In the next section, we will see how to rotate and resize coordinates and paths and perform node transformations.

Rotating, scaling, and slanting

A **rotation** means rotating a coordinate, node, or path around another coordinate or an axis. We will look at this first.

The following options can be used to rotate coordinates and paths:

- rotate is a value in degrees that rotates a coordinate or the coordinate system of a path around the origin. When used as an option for a node, the node's center is considered the origin of the rotation.
- rotate around takes an angle in degrees and a coordinate. This rotates around the given coordinate by that angle.

The following command draws a filled triangle that is rotated around the origin by 45 degrees:

```
\fill[orange, rotate=45] (0,1) -- (3,1) -- (2,2) --cycle;
```

You can compare it with the original triangle, filled with the color gray in the following figure:

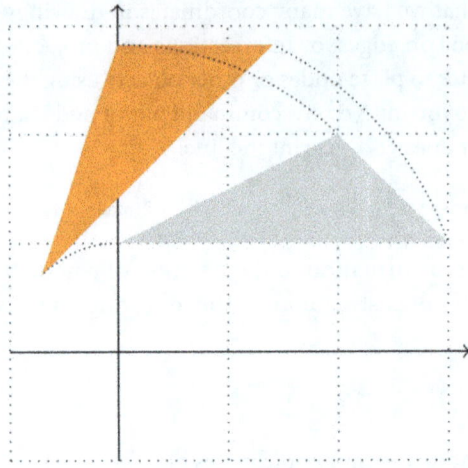

Figure 11.1 – Rotating a triangle around the origin

This command also rotates the triangle by 45 degrees, but around the coordinate (0,1) as the center of the rotation:

```
\fill[orange, rotate around={45:(0,1)}]
   (0,1) -- (3,1) -- (2,2) --cycle;
```

You can see the difference here:

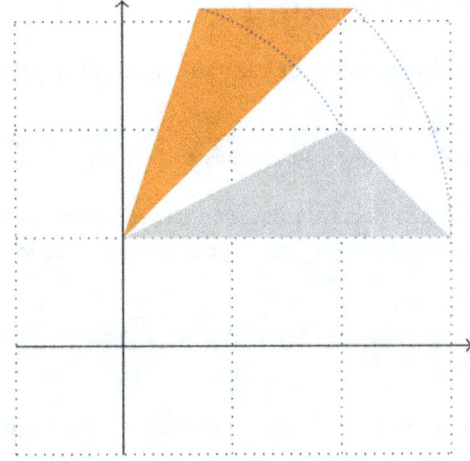

Figure 11.2 – Rotating a triangle around a point

You can also use a coordinate or node name, such as `rotate around={45:(A)}`. Remember that we had to use the curly braces to prevent the comma in a coordinate from being parsed as an option separator? Since we don't have a comma in node A, we can omit the braces and write `rotate around=45:(A)`.

In a three-dimensional coordinate system, as we used for *Figure 2.7*, you can rotate around each axis by an angle in degrees using one of the following options:

- `rotate around x` is a value in degrees for a counterclockwise rotation around the *x* axis. This is also called a **roll**.
- `rotate around y` is an angle for a counterclockwise rotation around the *y* axis. This is called a **pitch**.
- `rotate around z` is an angle for a counterclockwise rotation around the *z* axis. That has the name **yaw**.

Of course, we can use a combination of them. Roll, pitch, and yaw plus the three translations along the *x*, *y*, and *z* axes are the six degrees of freedom of an object moving in three-dimensional space. Those terms are used, for example, in aeronautics and physics.

Now let's look at resizing. We can do this by setting a scaling factor. Look back at *Figure 5.6*, the picture with smileys in different sizes; there, we already used scaling together with rotating. The following options are available:

- `scale` is a factor by which the coordinates are multiplied. So, values larger than 1 make them bigger, and values smaller than 1 make them smaller.
- `xscale` and `yscale` multiply only the coordinates' x value or y value. A negative factor flips the coordinates to the other side, so an `xscale` value of -1 is a flip, a mirroring at the *y* axis, similar to `yscale`.
- `scale around` takes a factor and a coordinate and resizes with the given coordinate as the center, similar to `rotate around`.

For example, the following command draws a double-sized triangle where the coordinate values are doubled:

```
\draw[scale=2] (0,1) -- (3,1) -- (2,2) --cycle;
```

So, after the transformation, TikZ draws a triangle with three corners (0,2), (6,2), and (4,4).

We can choose a point as the center of the transformation. For example, this way, we can resize the triangle by a factor of 2, with (0,1) as the center:

```
\fill[orange, scale around={2:(0,1)}]
    (0,1) -- (3,1) -- (2,2) --cycle;
```

That way, the transformed triangle still has the (0,1) corner and the two (6,1) and (4,3) corners, as you can see here:

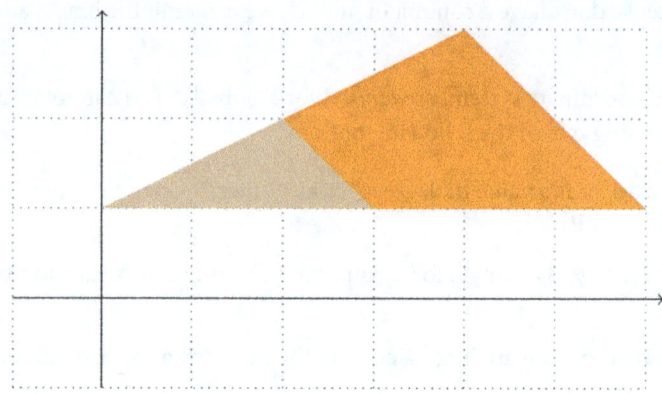

Figure 11.3 – A scaled triangle

You can also use a node name as the center of scaling.

Let's briefly look at what scaling by a negative factor means. As said previously, scaling x or y by -1 means flipping, mirroring at the *y* axis or *x* axis, respectively. A value of -2 means mirroring and then scaling by 2, and so on. The following code generates a mirrored image of a duck whom we already know and who has something to say:

```
\begin{scope}[xscale=-1, transform shape]
  \duck[laughing, speech={\tiny Oh a mirror!}]
\end{scope}
```

We get the following image, which you can compare with *Figure 9.1*:

Figure 11.4 – A mirrored scope

Note the `transform shape` option, which means that the node, including text, is also mirrored. We will examine that right after looking at applying multiple transformations.

We can combine scaling, rotating, and shifting. It's important to know that the order of transformation options can make a difference: transformations are applied in reverse order, like from right to left: the last given transformation is applied first. You may imagine it as transformation nesting. Remember school mathematics and compare it with nested functions f, g, and h in this way: f(g(h(x))) is evaluated from the inside to the outside, starting with h(x), which at the end appears evaluated from right to left.

But let's look at examples to be very clear. We will use simple nodes drawn by default at (0,0) when no position is given. This command rotates node A by 45 degrees and then moves it to the right by 2 cm, so the final position is (2cm,0):

```
\node[xshift=2cm, rotate=45] {A};
```

With reversed order, the following command moves node B by 2 cm to the right and then rotates it by 45 degrees around the origin (0,0), so the final position is (45:2 cm) in polar coordinates:

```
\node[rotate=45, xshift=2cm] {B};
```

In *Figure 11.5*, you can clearly see that swapping the order of transformations makes a difference. As an exercise, try to follow these transformations of nodes P and Q, now including mirroring:

```
\node[rotate=45, yshift=2cm, yscale=-1] {P};
\node[yscale=-1, yshift=2cm, rotate=45] {Q};
```

You can see their final positions in *Figure 11.5* as well:

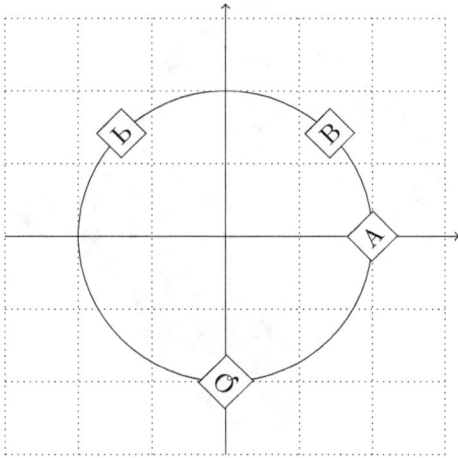

Figure 11.5 – Multiple transformations of nodes

You can use multiple transformations if you want to mirror objects along a line other than the *x* axis or the *y* axis. You can rotate the object to be aligned with an axis, shift it toward the axis, flip it with a scale value of -1, and then shift and rotate it back. The `tikz-ext` package provides a more convenient way. Using it, you can mirror an object at any line where you just specify two points that the line goes through. This additional package and its manual can be downloaded at https://ctan.org/pkg/tikz-ext.

In transformations, we can use macros as variables and calculate the values. Let's have a quick demonstration to practice `\foreach` loops and using scaling and rotating with a sequence of values.

Try to understand the following code:

```
\foreach \i in {90,85,...,5}
  \node[fill=black!\i, scale=\i, rotate=\i/2] {};
```

Here, the loop `\i` variable starts at 90 and goes down to 5 in steps of 5. The range is chosen because we use it as a blackness value, so we see that each drawing in the loop gets lighter. We can only use a math expression in a color value if we evaluate the loop variable as we did for *Figure 10.10*. We use the loop `\i` variable as the scaling factor, so our node, which has a rectangle shape by default, starts big and gets smaller. Because `scale=0` is as useless as `black!0`, which is just white, we let `\i` only decrease until 5.

While the loop variable goes down, the color gets lighter, and the node gets smaller, we rotate with an angle of half of the loop variable, starting at 45 degrees and decreasing over time.

The loop produces the following figure:

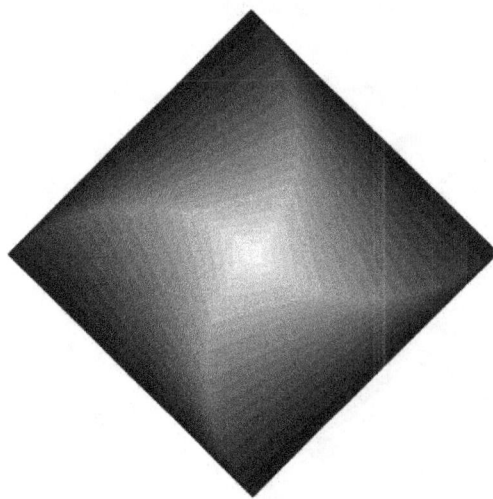

Figure 11.6 – Rotated and scaled squares

The node text is empty, so the rectangular shape is a square. What about scaling and rotating if the node contains text?

In the transformation section, the TikZ manual says, "*Scaling text is evil, rotating slightly less so.*" It's considered evil because text should not be scaled, but a larger font version should be used instead for the best quality. Because of this, the node text is not affected by transformations by default. However, you can enforce it. There are two ways:

- Use the transformation options within the node option list
- Add the `transform shape` option to the corresponding `\draw`, `scope`, or the `tikzpicture` environment

The second way is the easiest; we don't need to repeat scaling and rotating options. We will use scaled nodes in *Figure 11.8*. Let's look at slanting first.

The options are as follows:

- `xslant` slants the coordinates in the x direction, that is, horizontally, by a given value
- `yslant` does the same vertically, in the y direction, by the given value

A slant value of 0 has no effect, and a `yslant` value of 1 transforms (0,0) into (0,0), (1,0) into (1,1), (3,0) into (3,3), (1,1) into (1,2), and so on. A negative value slants it the other way around; we will see it in a minute.

Let's inspect the effect on a rectangle or, even better, on a grid. The following command slants a 3x3 grid by a factor of 0.5:

```
\draw[yslant=0.5] (0,0) grid +(3,3);
```

The grid now looks skewed like this:

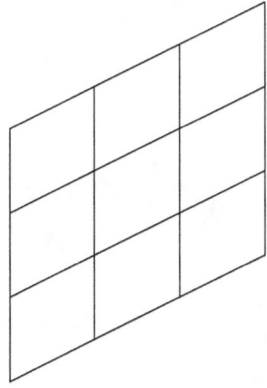

Figure 11.7 – A slanted 3x3 grid

Let's put a small drawing together where we can see the slant effects. At first, we draw such a slanted 3x3 grid from *Figure 11.7* again, just together with a 3x3 rectangle to get a shading effect with a color transition:

```
\draw[yslant=0.5,
   left color=gray!10, right color=gray!70]
   (3,-3) rectangle +(3,3)
   (3,-3)    grid    +(3,3);
```

Left of it, at (0,0), we draw such a grid with a negative yslant value, so it's skewed downward:

```
\draw[yslant=-0.5,
   left color=black!50, right color=gray!10]
   (0,0) rectangle +(3,3)
   (0,0)    grid    +(3,3);
```

Then, we draw a third pair, using a positive yslant value and adding a negative xslant value:

```
\draw[yslant=0.5, xslant=-1,
   bottom color=gray!10, top color=black!80]
   (3,0) rectangle +(3,3)
   (3,0)    grid    +(3,3);
```

That gives us three sides of a cube drawing. Before we look at that, let's also add slanted node texts. To get it big enough, we use a scale factor to demonstrate that scaling also works with node text when explicitly specified.

This creates our nodes:

```
\node[yslant=-0.5, scale=3.2] at (1.5,1.75) {TikZ};
\node[yslant= 0.5, scale=3.2] at (4.5,1.75) {Cube};
```

Now, compile and you get the following picture:

Figure 11.8 – A cube made from slanted grids

The lower three corners are, from left to right, at (0,0), (3,-1.5), and (6,0). You can see the slant effect works the same on text.

Apart from all these transformations, there's an option to preserve size and orientation. If you use rotating and scaling to get to a specific position for an object, but that option should only be placed there without rotation or a size change, then add the `shift only` option. In other words, the transformations will be applied to the position but won't change the object itself.

The following section will discuss how to treat a case when a transformation doesn't show the desired effect.

Transforming the canvas

Look at *Figure 4.1*, where we had an arrow between two nodes. The code for the arrow was the following:

```
\draw (tex) edge[->] (pdf);
```

When we want to draw a double arrow, a straightforward approach is to draw two such arrows and shift one up and one down. Let's do this, and add some rotation to practice our new skills:

```
\draw (tex) edge[->,yshift= 0.1mm, rotate= 4] (pdf);
\draw (tex) edge[->,yshift=-0.1mm, rotate=-4] (pdf);
```

If you compile, you may be surprised: both the shift and rotation don't have any effect; the arrow is the same in both cases.

In such a situation, we can transform the **canvas** instead of the coordinates. The canvas is our drawing area, like a sheet of paper, and a canvas transformation applies to everything: coordinates, text, line widths, everything. It happens on a lower level, with PDF or PostScript features, so we cannot track nodes or sizes at that time. Still, we can use it on a path to enforce a transformation. We can achieve this by putting our transformation options into a `transform canvas` option like the following:

```
\draw (tex) edge[->, transform canvas = {yshift= 0.1mm,
  rotate= 4}] (pdf);
\draw (tex) edge[->, transform canvas = {yshift=-0.1mm,
  rotate=-4}] (pdf);
```

With that change, *Figure 4.1* now gets two shifted and rotated arrows:

Figure 11.9 – Transformed arrows

Note, with a scaling transformation, the line width would also be scaled, which would not happen with a coordinate transformation.

Summary

Now you have the tools for shifting, rotating, and slanting coordinates, paths, and nodes. Using geometric transformations, you can now position objects and transform their orientation and shape in an advanced way.

In the next chapter, we will again focus on drawing. Specifically, we will learn about several ways to draw smooth curves that please the eye.

Further reading

The TikZ manual at `https://texdoc.org/pkg/tikz` covers transformations in *Part III* in the following sections:

- *Section 17.7, Transformations*, tells you how to transform nodes. You can find it online at `https://tikz.dev/tikz-shapes#sec-17.7`.

- *Section 25, Transformations*, deals with transformations in general. It is available online at `https://tikz.dev/tikz-transformations`.

- *Part IX, Section 99.4, Coordinate versus canvas transformations*, explains the difference between coordinate and canvas transformations. The online link is `https://tikz.dev/base-design#sec-99.4`.

You can read the `tikz-ext` package documentation at `https://texdoc.org/pkg/tikz-ext`.

12
Drawing Smooth Curves

In the first few chapters of this book, you learned about the TikZ tools for drawing lines, arrows, and shapes such as rectangles, circles, ellipses, and arcs. Now that you have advanced your TikZ skills, you are ready to tackle even more complex curves.

In this chapter, we will go through the following steps:

- Manually creating a smooth curve through chosen points
- Using a smooth plot to connect points
- Specifying cubic Bézier curves
- Using Bézier splines to connect given points
- Using the Hobby algorithm for smoothly connecting points

As we explore each method, we will compare the results of different methods with the same reference curve so you can see how different techniques affect the outcome. This chapter focuses on creating freehand-like drawings of nicely rounded curves without exact parameters.

By the end of this chapter, you will be able to draw easy curves just like by hand with a pencil – smooth and seamless without gaps, spikes, or corners. Furthermore, you will know how to take any curve without knowing how it's created and parameterized and generate a TikZ picture based on it.

Technical requirements

You can run and download the code of this chapter's examples at `https://tikz.org/chapter-12`. On GitHub, you can find it at

`https://github.com/PacktPublishing/LaTeX-graphics-with-TikZ/tree/main/12-drawing-smooth-curves`.

We will use the `spline` TikZ library, which you can download at https://github.com/stevecheckoway/tikzlibraryspline. Put it into your document folder or into your TeX distribution directory tree, where LaTeX can find it. Furthermore, we will use the `hobby` library included in regular LaTeX distributions.

Manually creating a smooth curve through chosen points

Our first goal is to draw a curve through several points that look round at any point. We will draw it similarly to a given curve as a second goal.

In the *LaTeX Cookbook* by Packt Publishing, in *Chapter 10, Advanced Mathematics*, there is a function plot that looks like the following:

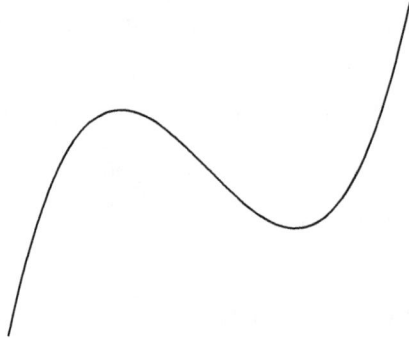

Figure 12.1 – A sample curve without coordinates or parameters

If you don't own the book, you can see that plot with code online at https://latex-cookbook.net/function-plot.

We will try to recreate this curve in the following steps:

1. We will identify the coordinates of a few points of the curve.
2. We will draw curve segments through these points to make it look like the original.
3. We will adjust each segment's start and end angle, as well as bending or the looseness of the curve, compile, look, and repeat until it looks as desired.

In the first step, we can include the source image as a regular image file, for example, called `curve.png`, by using `\includegraphics` in the `node` text, such as the following:

```
\node[opacity=0.5] {\includegraphics{curve}};
```

The opacity makes the image a lighter gray so that we can see our own drawing well in black. Then we add a grid as we did in previous chapters. This time, let's have two grids – one with big steps of 1 cm and a gray one with smaller steps of 0.2 cm – to estimate coordinate values more easily:

```
\draw[very thin, gray, step=.2] (-3,-3) grid (3,3);
\draw[step=1] (-3,-3) grid (3,3);
\draw[->] (-3,0) -- (3,0) node[right] {x};
\draw[->] (0,-3) -- (0,3) node[above] {y};
```

Now we have a nice grid on top of the curve to start identifying coordinate values:

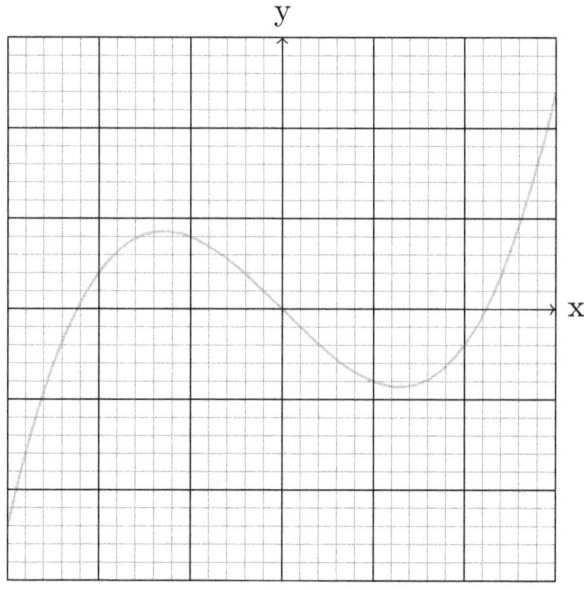

Figure 12.2 – A curve with a grid

By looking at the 1 cm grid and the smaller 2 mm grid, we can identify the coordinates of the curve's first and last points and a few points in between. We can draw the points in a \foreach loop as we learned in the previous chapter:

```
\foreach \x/\y in { -3/-2.4, -2/0.4, -0.4/0.4,
    0.4/-0.4, 2/-0.4, 3/2.4 }
        \fill (\x,\y) circle (0.6mm);
```

If we see that the points may be slightly off the curve, we could adjust their coordinate values a bit until we are satisfied and they are nicely on our original curve, as we see here:

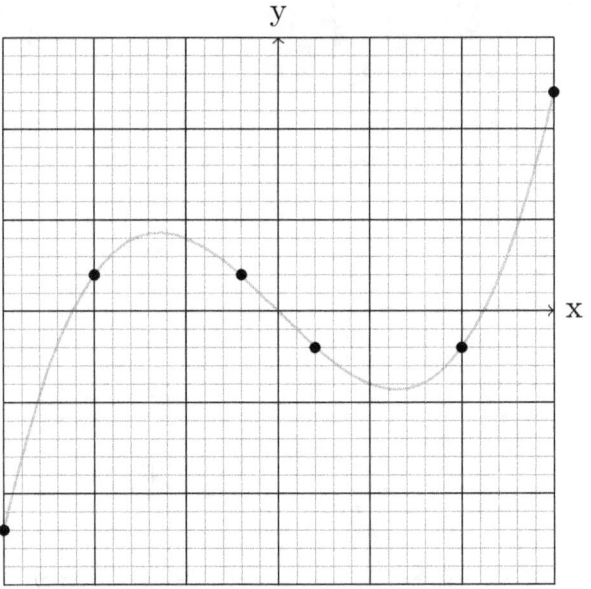

Figure 12.3 – Manually selected points on a curve

Now comes the exciting part. We will connect the points using the to operation that we introduced in *Chapter 4*. This draws straight lines by default; however, we can choose the start and end angle of each segment, and a looseness for fine-tuning, like in this segment, leaving the start point at 60 degrees, coming in at the end point at 135 degrees, with a looseness of 1.2:

```
\draw ... to[out=60, in=135, looseness=1.2] ...
```

The smart point here is to choose every point's incoming angle; the outgoing angle should then differ from the incoming angle by 180 degrees. This ensures that the curve runs through a point like a tangent.

We estimate the angles, compile, and adjust if we see it a bit off until we have good values. This is what I came up with:

```
\draw (-3,-2.4) to[out=77, in=240] (-2,0.4)
  to[out=60,  in=135, looseness=1.2] (-0.4,0.4)
  to[out=-45, in=135] (0.4,-0.4)
  to[out=-45, in=-120, looseness=1.2] (2,-0.4)
  to[out=60,  in=257] (3,2.4);
```

Admittedly, to fit the original curve, it took me several tries to get the values right so it looks like the original curve. If you want to draw a smooth curve through some points, it's enough to remember to keep the 180 degrees difference between incoming and outgoing.

Without the grid, this is now our curve, using the preceding \draw command with the to operation:

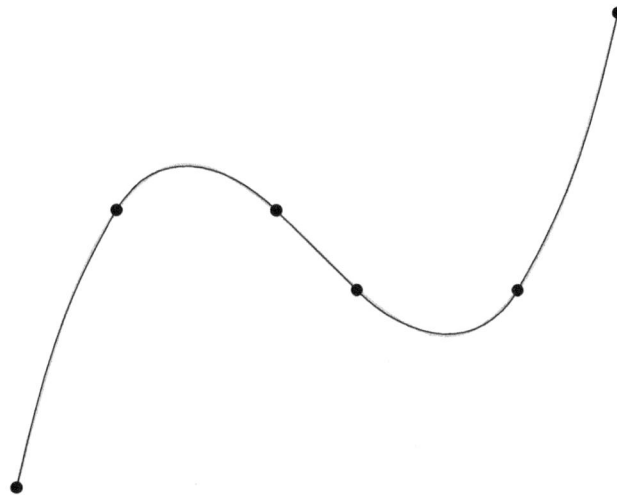

Figure 12.4 – A smooth path through points

That resembles our original curve close enough and is indeed very smooth.

If we want to draw a more complex curve, we may need more points. In that case, it can be too laborious to get all the in and out angles right. Let's look at another approach in the next section.

Using a smooth plot to connect points

TikZ can plot functions for us, either using a set of coordinate values or a mathematical parametrization. We will learn a lot about plotting in the next chapter; for now, let's have just a first quick look at coordinate plots.

We will work with the set of points from the previous section. As said, we may need more points for more accuracy, so let's look at *Figure 12.2* again and choose two additional coordinates on the curve, such as two peak values (-1.3, 0.86) and (1.3, -0.86). TikZ can do a simple plot through all those points by the plot operation as follows:

```
\draw plot coordinates {
    (-3,-2.4)  (-2,0.4)  (-1.3,0.86)  (-0.4,0.4)
    (0.4,-0.4)  (1.3,-0.86)  (2,-0.4)  (3,2.4)  };
```

These commands generate a sequence of straight linear segments:

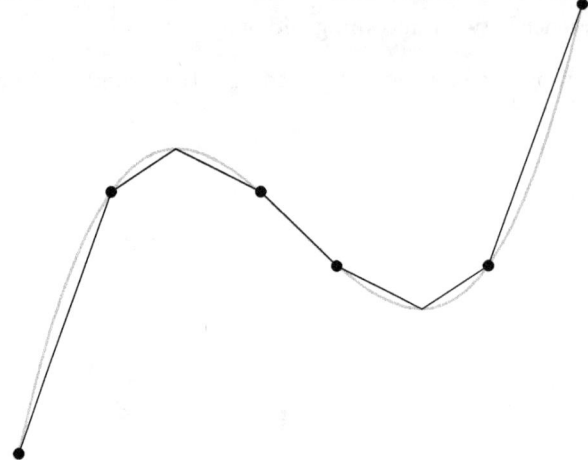

Figure 12.5 – A plot through given coordinates

That's not smooth yet, but add the smooth keyword as an option to plot:

```
\draw plot[smooth] coordinates {
    (-3,-2.4)   (-2,0.4)    (-1.3,0.86)  (-0.4,0.4)
    (0.4,-0.4)  (1.3,-0.86) (2,-0.4)     (3,2.4) };
```

Compile again, and we get a nice smooth curve as follows:

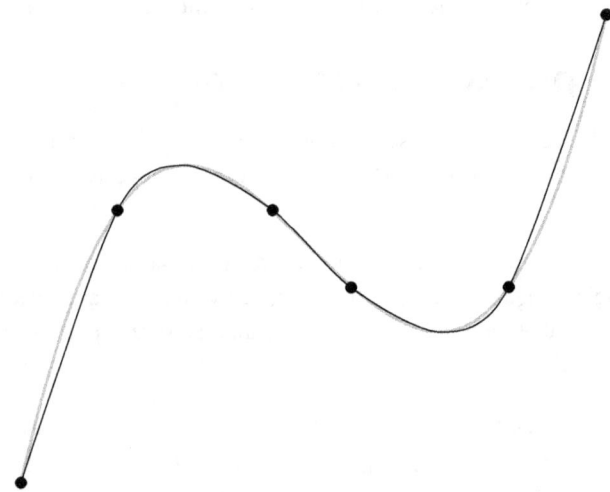

Figure 12.6 – A smoothened plot through the given coordinates

Now, TikZ has taken care of the transition points. Just the first and the last segment look straight, which can be okay or not desired. In any case, you can choose a few more points where the curve shall pass through to make it fit as you want. At least you don't need to think about incoming and outgoing angles.

The next section will show another way to define curves with a few parameters.

Specifying cubic Bézier curves

In the previous section, we saw that linear segments are not a good curve approximation. We could use quadratic curves and parabola segments to build rounder curves. Even better and more flexible are cubic curves. In computer graphics, so-called **Bézier curves** are used to approximate other curves, which are polynomial curves. Cubic Bézier curves are good enough and already complicated enough.

At the end of the chapter, in the *Further reading* section, you will get links to websites where you can read about the mathematics of Bézier curves. Here, we will look at them in a basic user approach, focusing only on the cubic curves that TikZ supports.

In TikZ, we can declare a curve from coordinates A to B with control points P and Q in the following way:

```
\draw (A) .. controls (P) and (Q) .. (B);
```

The curve starts in A in the direction toward P, which means that the line A to P is a tangent in A. Then, it ends in B coming from the direction of Q, meaning that the line Q to B is tangent. P and Q are not on the path itself. The higher the distance between P and Q is, the more the curve turns toward P and later toward Q, which means the higher the amplitude of the curve. It's like P and Q pull the curve to them, stronger, if they are far away.

In our case, A and B are the curve's very left and right points. It takes a while to experiment with values for P and Q, but this choice comes close to the original curve:

```
\draw[dotted] (-3,-2.4)
  .. controls (-1,6.4) and (1,-6.4) .. (3,2.4);
```

Figure 12.7 shows the control points and the fact that they are not arbitrarily chosen. As said, the connection to the corresponding end point shall be tangent.

Using relative polar coordinates can make it clearer. As we saw in the `to` operation approach in the first section of this chapter, the angle at the start is about 77 degrees, and the angle at the end is about 257 degrees. We can use this and set the polar distance high enough here to 9:

```
\draw[thick] (-3,-2.4)
   .. controls +(77:9) and +(257:9) .. (3,2.4);
```

In both cases, this is the picture we get, with P and Q and help lines displayed in gray:

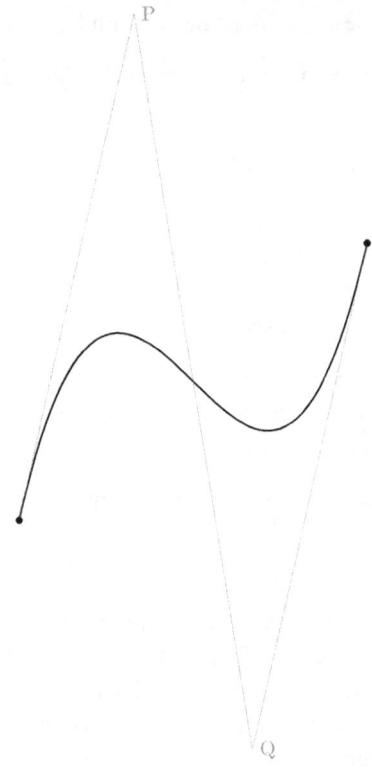

Figure 12.7 – A Bézier curve with control points

How does it happen that we can define and draw that curve using only four control points? We are lucky here, as the curve itself is a cubic curve, `x^3/5-x`, and not of a higher degree or trigonometric or very random. So, it's not super-hard to approximate.

For more complex curves, we may need to use several Bézier curves. In the next section, we will see how easy this can be.

Using Bézier splines to connect given points

The previous methods may be too laborious when we want to create a more complex curve defined by many points. It's hard enough to find the control points for a desired Bézier curve by trial and error. And if we need a series of Bézier curve segments, called **splines**, it could be a nightmare.

Luckily, there's the `spline` library. We can load it as follows:

```
\usetikzlibrary{spline}
```

Then we can specify end points as before and a spline through a set of coordinates that shall be passed through:

```
\draw[thick]  (-3,-2.4)
   to[spline through={(-1.3,0.86)(1.3,-0.86)}] (3,2.4);
```

The library creates a path consisting of Bézier curve segments. It looks as follows, where I additionally plotted the used control points in gray:

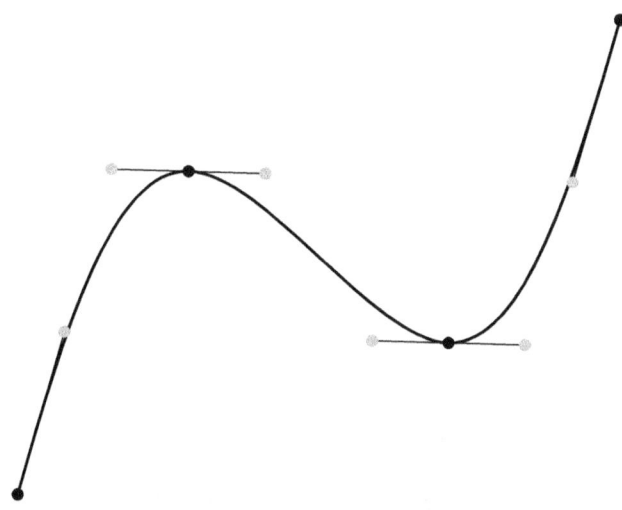

Figure 12.8 – A curve with Bézier splines

Here, you can see that, for neighbor splines, the end and start control points are on a tangent line.

In the next section, we will see a similar approach to connect curve segments with low overall curvature.

Using the Hobby algorithm for smoothly connecting points

John Hobby, the creator of the **MetaPost** graphic language, developed an algorithm for drawing a curve through a given set of points. That's similar to the previous section's approach because it internally creates a list of cubic Bézier curves. The curves are parameterized in order to be joined together very smoothly. This provides very pleasing results. It's not about perfect approximation; it's about very smooth curves.

For us, it's just another syntax with a different result. First, load the `hobby` library:

```
\usetikzlibrary{hobby}
```

Now, we set up a plot with start and end coordinates and two intermediate coordinates. We will take a few coordinates we used for *Figure 12.5* to get a comparable result. The main difference is that we use hobby as a plot option:

```
\draw plot[hobby] coordinates { (-3,-2.4) (-1.3,0.86)
   (1.3,-0.86) (3,2.4)};
```

Compile it, and you get a curve that's amazingly round and smooth:

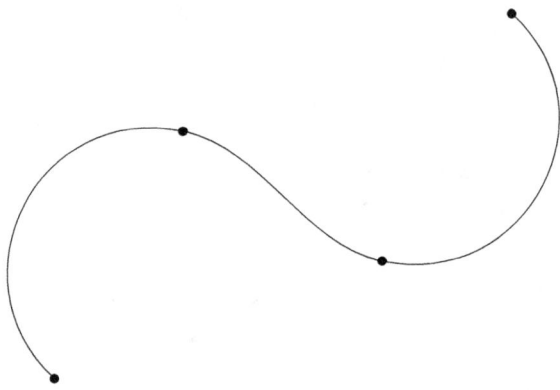

Figure 12.9 – A Hobby curve

Apart from plotting, the hobby library provides a more straightforward syntax, using the to operation, which you know about from *Chapter 4, Drawing Edges and Arrows*. We draw with a start point and an end point and specify where the path shall go through using a curve through = {coordinates} option. It's immediately clear when you see an example. The following code is equivalent to the previous plot:

```
\draw (-3,-2.4)
    to[curve through = {(-1.3,0.86) (1.3,-0.86)}]
   (3,2.4);
```

You can decide which syntax you prefer.

To get a smoothly closed curve, we can add the closed option like this:

```
to[closed, curve through = {(-1.3,0.86) (1.3,-0.86)}]
```

It's closing the path like this:

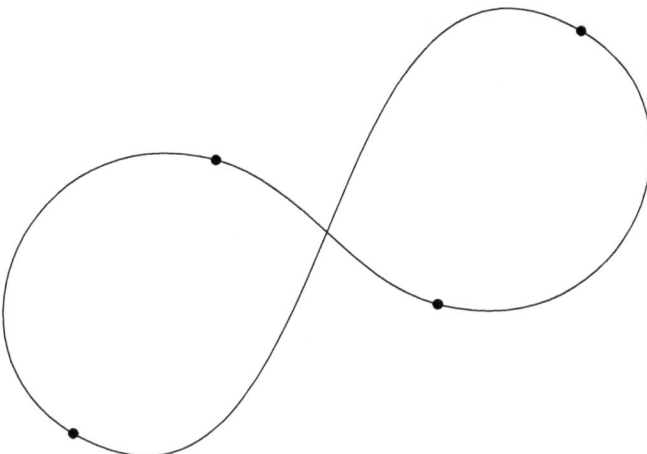

Figure 12.10 – A closed smooth curve

If you look closely at the start and the end of the hobby curve in *Figure 12.9*, you can see that the first and the last segment are drawn like circular arcs. You can change this default behavior by explicitly stating out and in angles as follows:

```
\draw (-3,-2.4) to[out angle=80, in angle=260,
   curve through = {(-1.3,0.86) (1.3,-0.86)}] (3,2.4);
```

This changes *Figure 12.9* in the following way:

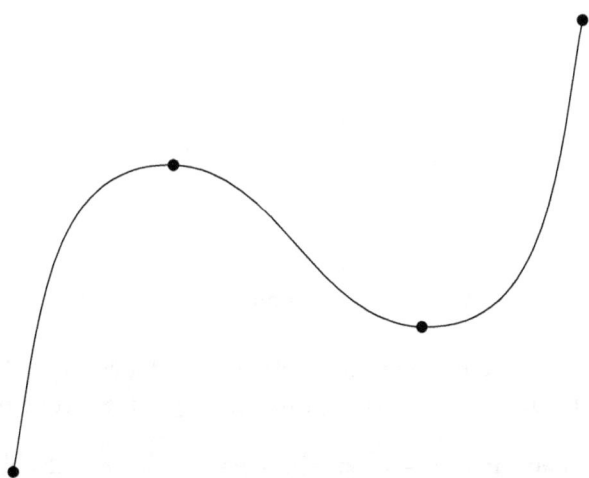

Figure 12.11 – A curve with a specific start and end angle

By changing the start angle to 80 degrees and the end angle to 260 degrees, we made the curve look less round and more similar to *Figure 12.1*, where we started. Note that we only stated a start point, an end point, and two points in between, to get a fantastic smooth curve through them.

The `hobby` package supports several options called `curl` and `tension` to fine-tune bending and looseness. If you would like to learn more about those fine details, please take a look at the `hobby` manual. For now, let's play more with the default curve shapes.

We can create more intricate curves with more points. Let's say we want to create some random blob curve that looks nice. We can choose some points in the xy plane where the curve passes. Here, we draw some:

```
\foreach \c in {(0,0),(-1,-2),(-2,-1),(-1,0),
    (-1,2),(0,1),(2,1)} \fill \c circle (0.5mm);
```

Their position in the xy plane is as follows:

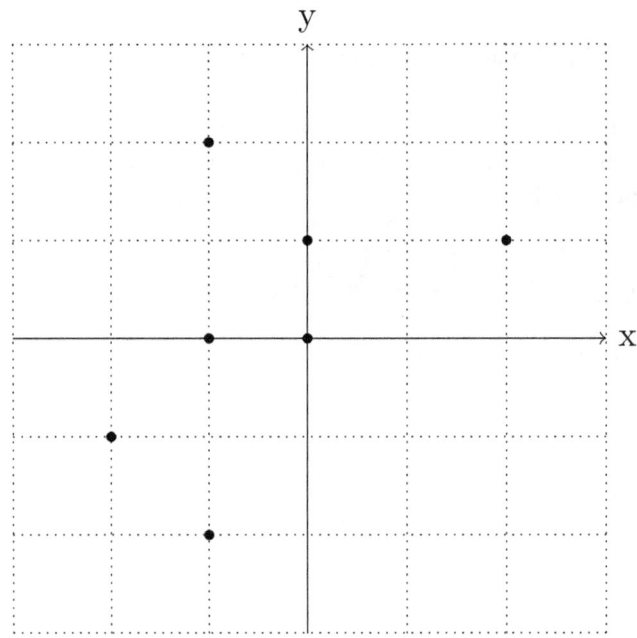

Figure 12.12 – Points on the plane

Now, the challenge is drawing a smooth curve through all the points. We can try to do it by hand on paper first. Drawing with any method looks difficult, except with `hobby` – it's a breeze. Try the following line:

```
\draw[thick, fill=gray] (0,0) to[closed, curve through =
    { (-1,-2) (-2,-1) (-1,0) (-1,2) (0,1) }] (2,1);
```

This is the shape we get:

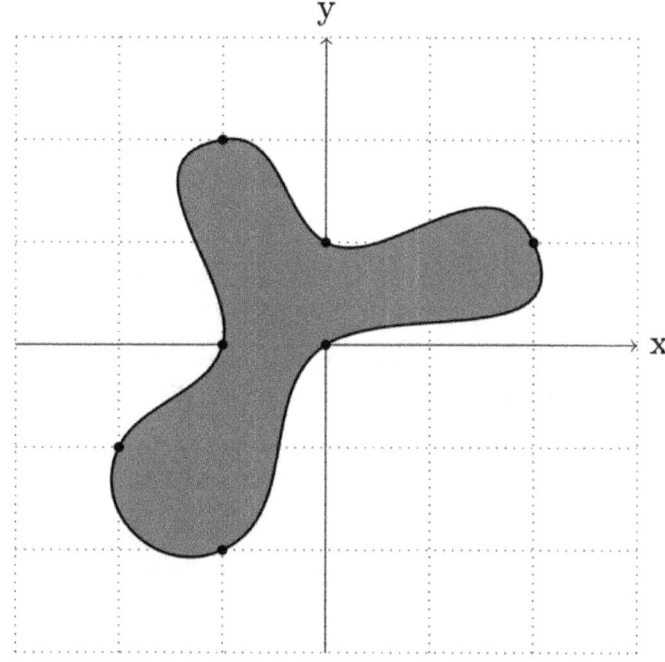

Figure 12.13 – Points on the plane

That's a fantastic smooth way of connecting points! For fine-tuning, you can move the points around a bit, recompile, and repeat a few times until you get a shape you like.

Summary

In this chapter, you learned how to approximate curves that are complex or somehow randomly chosen or where we don't know the mathematical parametrization.

And more importantly, you learned how to create smooth, elegant curves through a set of points you chose.

In the next chapter, we will talk much more about plotting functions, but at that time, we will deal with explicit mathematics.

Further reading

In the TikZ manual at `https://texdoc.org/pkg/tikz`, our topics are covered in *Part III* in the following sections:

- *Section 14.3, The Curve-To Operation* is about the curve syntax using control points. It's online at `https://tikz.dev/tikz-paths#sec-14.3`.
- *Section 22, Plots of Functions* describes TikZ's basic plotting syntax, including the `smooth` option. You can read it online at `https://tikz.dev/tikz-plots`.

`https://en.wikipedia.org/wiki/B%C3%A9zier_curve` in Wikipedia is a good starting point for reading about Bézier curves and finding further documents.

`http://weitz.de/hobby` is a JavaScript demonstration of the Hobby algorithm versus cubic splines, where you can define and move points by simple mouse clicks in a drawing.

The `hobby` library is explained at `https://texdoc.org/pkg/hobby`. For even more background, you can read the MetaPost manual at `https://texdoc.org/pkg/metapost`, particularly *Section 4.2, Specifying Direction, Tension, and Curl*.

13
Plotting in 2D and 3D

Whether you're a scientist, analyst, engineer, teacher, or student, you know that proper visualization is vital to understanding your data.

If you decide to showcase your data using diagrams such as line charts, bar charts, or pie charts, then *Chapter 14, Drawing Diagrams*, will cover you.

If you want to visualize your data in a coordinate system in LaTeX, then the current chapter is the right place.

In this chapter, we'll be covering the following topics:

- Introducing plotting
- Creating and customizing Cartesian axes, ticks, and labels
- Using plotting commands and options
- Filling the area between plots
- Calculating plot intersections
- Adding a legend
- Using the polar coordinate system
- Parametric plotting
- Plotting in three dimensions

After studying this chapter, you will be able to easily plot datasets and mathematical functions in a scientific and technical context.

Technical requirements

This chapter's examples are at `https://tikz.org/chapter-13`. On GitHub, they are at `https://github.com/PacktPublishing/LaTeX-graphics-with-TikZ/tree/main/13-plotting`.

The `pgfplots` package is crucial here and must be installed in your LaTeX distribution. `pgfplots` is built on the pgf/TikZ packages, so you must have it installed and loaded in your document. If you don't load pgf/TikZ explicitly, `pgfplots` will load it automatically. If you use `TikZ.org`, Overleaf, or `TeXlive.net`, `pgfplots` is already included. We will also use the `pgfplots` libraries `colormaps`, `fillbetween`, and `polar`, which are bundled with `pgfplots`.

Introducing plotting

Let's begin by discussing and briefly assessing the available options for plotting, both in the commercial software market and the open source community.

Several commercial software options are available, such as **Mathematica** and **Matlab**, as well as free and open source software such as **GNU Octave**, **R**, **GNUplot**, and **Python** with **Matplotlib**, for example. Any of these programs can be used to generate plots and export them as images for inclusion in your LaTeX document. However, there are some drawbacks to using third-party software:

- You need to install and maintain another software installation, you may have to pay for installation and updates, and you will be dependent on the software's functionality on a specific computer.
- Imported images may appear blurry or pixelated if exported in bitmap formats such as PNG or JPG. If possible, export them as PDF images to obtain a scalable image.
- Font types and the sizes of labels and numbers may differ from those in your LaTeX document.
- Formulas, mathematical symbols, arrows, and line widths may differ significantly from those in LaTeX.

MetaPost and **Asymptote** are programming languages utilized for plotting and integrating appropriately with LaTeX. In particular, Asymptote is extremely powerful and provides excellent 3D capabilities. However, both of these options require learning a new syntax. Similarly, **PStricks** can also be used for plotting; however, it's not comparable to the TikZ syntax, with which you are already proficient.

Furthermore, there are several advantages to producing plots directly in LaTeX using TikZ, including the following:

- TikZ and its plotting packages are immediately available in a complete LaTeX installation
- TikZ generates sharp, high-quality plots
- You get a consistent drawing with LaTeX-rendered formulas and symbols that use your document's fonts

- You can define document-wide styles for plots and customize them in your preamble without repetition in the document
- Many features and styles have been designed to work seamlessly in the LaTeX environment

In *Chapter 12, Drawing Smooth Curves*, we met the TikZ `\plot` path operation. It can plot coordinates and even mathematical functions for us. It's okay for a quick plot without axes, a grid, or sophisticated features. Furthermore, TikZ recently added the `datavisualization` libraries, which are somewhat academic, sophisticated, and ambitious. That's excellent!

Before the introduction of the `datavisualization` libraries, Christian Feuersänger developed the `pgfplots` package on top of TikZ. It grew fast and became rich in 2D and 3D features, and has a large user base and numerous examples on the internet. This chapter will focus on `pgfplots` as this is an established and proven package.

The `pgfplots` package has an excellent reference manual. It is vast, with about 600 pages, very detailed, with a lot of examples. This chapter will give you a fast-paced introduction with some examples. Once you read through this chapter and start plotting data and functions yourself, explore the `pgfplots` manual to use its complete reference of customization options. Here, we focus on the most commonly used options and features. Specifically, we selected exciting topics that go beyond pure plotting.

In addition, check out the TikZ and `pgfplots` galleries listed at the end of this chapter. They contain many examples with the complete code, giving you starting points and further insights into designing plots.

To use `pgfplots`, we always need to load the package in the document preamble:

```
\usepackage{pgfplots}
```

`pgfplots` has a remarkable way of ensuring backward compatibility. Due to fast development, many additions, and changes over time, `pgfplots` introduced a compatibility setting. For example, if you use `pgfplots` version 1.18 from 2023, as I did for this book, you should add this statement to your preamble:

```
\pgfplotsset{compat=1.18}
```

This ensures that in later versions, such as 2.0, your plots will look the same as in version 1.18, no matter what has been changed in 2.0.

You could use `\pgfplotsset{compat=newest}` to always use the newest features, even after package or distribution updates. However, when you later compile with updated versions, your plots may be rendered differently, so the `newest` setting is discouraged, but is provided nevertheless.

The remainder of this chapter will explore many of the `pgfplots` features. We will plot small datasets and mathematical functions. If you need to plot larger datasets, such as the results of lab experiments, the next chapter will cover plotting data from external files, where we will use that feature for rendering charts.

Plotting in 2D and 3D

We will continue with the fundamental task of creating and fine-tuning coordinate axes.

Creating and customizing Cartesian axes, ticks, and labels

In *Chapter 12, Drawing Smooth Curves*, we identified and plotted a few points with self-made axes and a grid. We will plot them now using `pgfplots` to get a first glance at the syntax.

Take a look at this code, which you can download from GitHub or the Chapter 13 page on `TikZ.org`:

```
\documentclass{article}
\usepackage{pgfplots}
\pgfplotsset{compat=1.18}
\begin{document}
\begin{tikzpicture}
  \begin{axis}[grid]
    \addplot[only marks] coordinates
      { (-3,-2.4)   (-2,0.4)   (-0.4,0.4)
        (0.4,-0.4)  (2,-0.4)   (3,2.4) };
  \end{axis}
\end{tikzpicture}
\end{document}
```

By compiling this document, we get the following output:

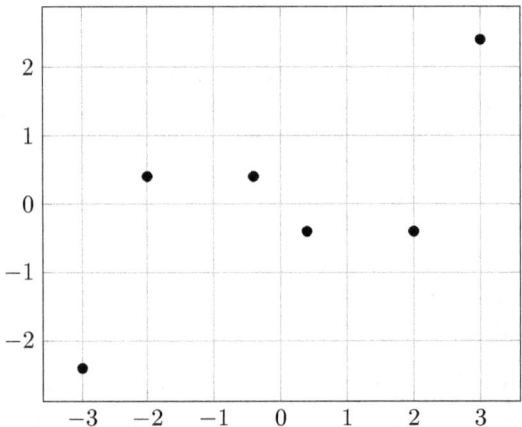

Figure 13.1 – Plotting coordinates

At first, we load the `pgfplots` package and set version `1.18` for compatibility. As `pgfplots` uses TikZ, every plot must be in a `tikzpicture` environment.

A plot is done within an `axis` environment. This makes sense, as the axis, including the coordinate ranges, defines the playing field. The `axis` environment draws our coordinate axes, and the `grid` option includes a rectangular grid. The `\addplot` command is used for generating a plot. The `only marks` option is used to determine that only marks for the coordinates are plotted without connecting lines.

For the rest of the chapter, we will omit the `tikzpicture` environment. You know well that it encloses a plot, so we don't need to be repetitive.

We can set various options for axes and plots, of which there are a few different types. We look at axis types first.

Understanding axis environments

We can choose from the following axis environments:

- `axis` creates a linear axis like that used in the Cartesian coordinate system. We will use it here most of the time. It can also be used for 3D drawing.
- `semilogxaxis` enables logarithmic scaling of the x-axis.
- `semilogyaxis` enables logarithmic scaling of the y-axis.
- `loglogaxis` does logarithmic scaling for both the x-axis and y-axis.
- `polaraxis` creates a circular axis with polar coordinates. We will look at it in the *Using the polar coordinate system* section later in this chapter. It requires the `polar` library.

In this chapter, we will use the `axis` and `polaraxis` environments to cover the main concepts. If you need logarithmic axes, please refer to the `pgfplots` manual.

By default, the axis has a box shape, and ticks and labels are placed along the box sides. This is often used in a scientific context, such as for plotting a lot of experimental data.

You may remember that "cross-axis" from school and university, where the x-axis and y-axis are centered, so they intersect at the origin. This can be achieved using the `axis lines=center` option. Let's look at this quickly while trying the next plot type, where we give a function to the `\addplot` command:

```
\begin{axis}[axis lines=center]
  \addplot[thick, samples=80, smooth, domain=-3:3]
    {x^3/5 - x};
\end{axis}
```

This plots a cubic function, which is a third-degree polynomial curve, with the centered axes that we know from school geometry and analysis:

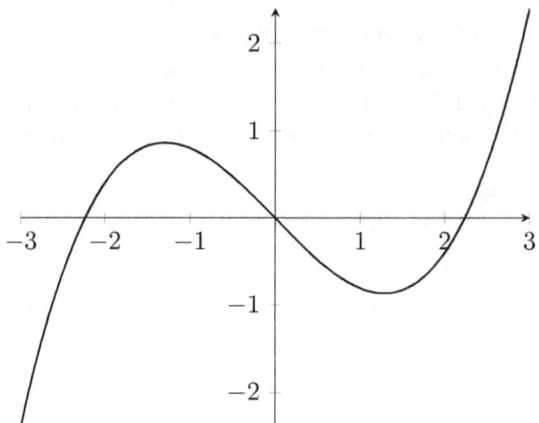

Figure 13.2 – A cubic plot with a centered x-axis and y-axis

That curve looks familiar: that's the function we tried to "hand-draw" as an example in the previous chapter.

When plotting functions, centered axes have advantages: you can immediately see when a function crosses zero, and you can better see symmetries and the position of a function on a plane. That's why we continue using it here for 2D plots.

We plotted the cubic function as a smooth, connected thick curve with 80 sample points. The `\addplot` options shall be our topic in the next section. For now, we continue with the axes.

For `axis line`, we can choose the following values:

- `box` is the default and draws a rectangle with ticks and labels on the outside.
- `center` and `middle` are synonymous and do the same; they draw centered axes intersecting at the origin (0,0). If the origin is not in the range of values of the axis, it draws the axis line at the lower side of the axis.
- `none` hides the axes; no axis is drawn. That's for when you want pure focus on a plot without distracting lines or labels.

You can choose axis lines individually, with the same values as we saw for `axis line` and additional styles, such as the following:

- `axis x line` can also be `top`, drawn at the maximum y value, or `bottom`, drawn at the minimum y value. In the case of `box`, both are drawn.
- `axis y line` can be `left`, drawn at the minimum x value, or `right`, drawn at the maximum x value. Again, with `box`, both are drawn. The same applies to `axis z line` in the case of 3D drawings.

`pgfplots` determines size and scaling automatically. In the case of *Figure 13.1*, the x and y ranges for the axes were evident since the plot was based on specific coordinate values. In *Figure 13.2*, where we had a mathematical function, we had to define the x range, the function's **domain**. By setting `domain=-3:3`, we got a range from x=-3 to x=3. The y domain was automatically calculated.

`pgfplots` calculates as much as possible and works with meaningful defaults, as you saw in our first examples. However, you can set the axis limits as you like:

- `xmin`, `xmax` define the minimum and maximum value of the *x*-axis.
- `ymin`, `ymax` set the minimum and maximum value of the *y*-axis.
- `zmin`, `zmax` do the same for the *z*-axis in a 3D plot.

They don't have to be the same as the domain of a plot.

When the x and y range are significantly different, one will be scaled. If you don't like this, use the `axis equal` option; then, a y unit will be the same size as an x unit on the axis. If necessary, the image will be enlarged to keep the aspect ratio of the whole plot. We will use this in the upcoming figures. Use the `axis equal image` option if you want `pgfplots` to not enlarge the range while enforcing equal x and y unit sizes. The latter option results in smaller images. You can see this in *Figure 13.4* in the next section.

In general, you can change the size and aspect ratio by using the following keys:

- `width` can be any TeX dimension, such as `width=6cm` or `width=\textwidth`
- `height` can also be set as TeX dimension, such as `height=4cm` or `height=0.25\textheight`

If you choose either `width` or `height`, the other one will be calculated to keep the aspect ratio. If you set both, the aspect ratio will change.

Customizing ticks and labels

You can customize the **ticks** on the axes. `pgfplots` chooses suitable values by default, as you can see in *Figure 13.1* and *Figure 13.2*. You can override the numeric distance between ticks using the `xtick distance`, `ytick distance`, and `ztick distance` axis options. You can set them in this way:

```
\begin{axis}[xtick distance=2, ytick distance=0.5]
   ...
\end{axis}
```

Here, we get *x*-axis ticks in a sequence such as -4, -2, 0, 2, 4, 6, ..., while the *y*-axis ticks will appear as -1, -0.5, 0, 0.5, 1, 1.5, and so on.

Note that in the following, for each x option, there exist y and z options of the same kind, even if I don't mention them all each time.

In the same way, we can set other tick options. If you want to remove the *x*-axis ticks altogether, set `xtick=\empty`. This works in the same way for `ytick` and `ztick`. For those three options, you can also choose specific values as a list in braces, such as `xtick={1, 2, 8, 10}`, which will give you only those chosen ticks at the exact corresponding axis location.

The `data` key, as in `ytick=data`, generates tick marks at every coordinate of the first plot.

You can also have so-called **minor ticks**, smaller ticks between the normal ones. Just give a number to `minor tick num` for all axes, `minor x tick num` for the *x*-axis, and so on, and that number of ticks will be printed equidistantly between the regular ticks. For example, with the `minor tick num=3` axis option, *Figure 13.2* changes as follows:

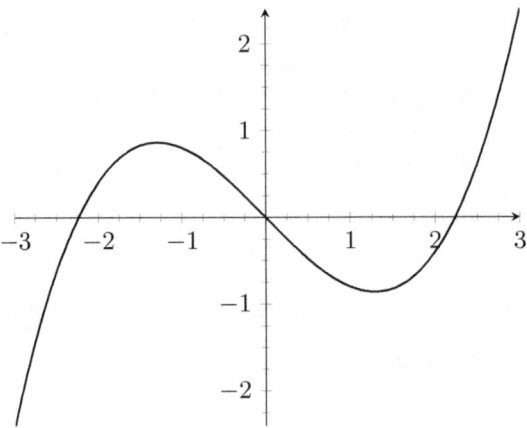

Figure 13.3 – Axes with minor ticks

If the default tick labels look too big, you can customize elements of the style such as the size and color. Use `tick label style = { ... }` for any adjustments. We will use it in the next drawing.

You can choose your own labels instead of the default numbers, for example, to have better-looking fractions. To do this, set, for instance, `xticklabels = { ... }`, and within the braces, use any list of values, symbols, or LaTeX commands that matches the number of your x ticks. You can do the same for y and z, if needed.

To see such options in a picture, let's now have a quartic function plot, a fourth-degree polynomial curve. But now we want to do it with our own customized ticks. We want to use fractions in the labels, and we would like to have smaller tick labels because otherwise, the fractions look too big.

We will use the following code; the tick options are highlighted:

```
\begin{axis}[axis lines = middle, axis equal image,
  domain = -1.25:1.25, y domain = 0:1.25,
  ymax = 1.2,
  tick label style = {font=\scriptsize},
  xtick = {-1, -0.5, 0.5, 1},
  xticklabels = {-1, $-\frac{1}{2}$, $\frac{1}{2}$, 1},
  ytick = {0.25, 0.5, 0.75},
  yticklabels = {$\frac{1}{4}$, $\frac{1}{2}$,
     $\frac{3}{4}$} ]
  \addplot { (x^2-1)^2 };
\end{axis}
```

This gives us axes with much nicer LaTeX style labels:

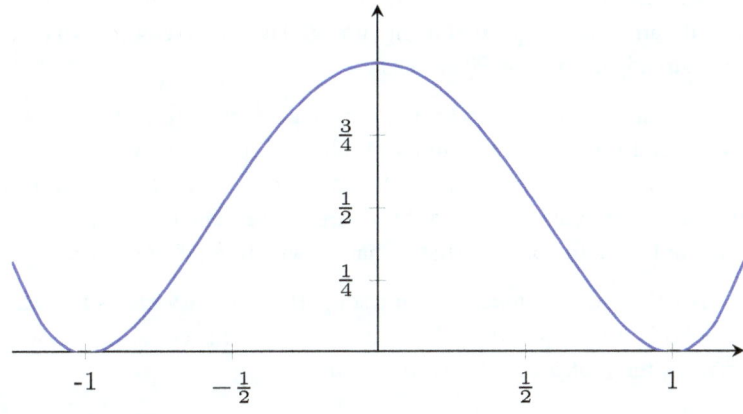

Figure 13.4 – Customized ticks

For even more fine-tuning, take a look into the pgfplots manual; in *Section 4.15, Tick Options*, you can find even more about tick positioning, shifting, scaling, and alignment. We will continue with a more general look at plot design now.

Using plotting commands and options

We already encountered the most important command, which is \addplot. You may have noticed that when we used \addplot with options, the color of the plot was black. When we did not use options, it was blue.

The reason is that a so-called **cycle list** contains the color and marker style for plots. So, by default, the first plot in a drawing would be blue, the second would be red, and the third would be green color. We leave the details of this to the `pgfplots` manual so we understand how the coloring happens.

So, when we use `\addplot[color=yellow, ...]`, the options provided will replace the default options.

The `\addplot+` command, however, appends the given options to the default options. We won't use it here, but it's good to know for when you see it used online and when you want to use the pre-defined `cycle list` of blue, red, and green for the first three plots in a diagram.

For both commands, we have three variants:

- `\addplot coordinates {...}` plots a set of coordinates, which we did for *Figure 13.1*.
- `\addplot table {...}` plots data from a table, either given as a filename or inline text, in the curly braces. We will cover this on the book's website https://tikz.org/plotting-data, together with sample data files.
- `\addplot {<math expression>}` is the most interesting for us here: the math expression will be evaluated, sampled, and plotted using the pgf/TikZ math engine. No external program is required. We already used it for *Figure 13.2*.
- `\addplot (<x math expression>, <y math expression>)` is for parametric plots where you can use a math expression for both the *x* coordinate value and the *y* coordinate value of the points in the plot. We will use this in the *Parametric plotting* section later in this chapter. Since math expressions may contain parentheses, enclosing them in curly braces is better to avoid confusing the parser. That's what we will do for *Figure 13.12*.
- `\addplot3` is for three-dimensional plots and supports the same variants as the aforementioned `\addplot`. `\addplot3` works in the same way. We will see examples of this in the last section of this chapter, *Plotting in three dimensions*.

We can always give style options in square brackets as usual, as exhibited in this chapter's examples.

There's also a command to set plot styles globally. Look at this:

```
\pgfplotsset{every axis plot post/.append style =
  {samples=80, smooth, thick, black, mark=none} }
```

This defines 80 sample points per plot, smoothing the curve and making it thick, black, and without markers. Once you write this, it applies to all your plots. However, you can override it at any time by placing another `every axis plot post/.append style` setting as an option to an axis environment or a single `\addplot` command.

In our other 2D function drawings, we will continue using this style.

The next section will show how to fill the area below a continuous plot or between two plots.

Filling the area between plots

In *Chapter 7, Filling, Clipping, and Shading*, we dealt with filling areas enclosed by TikZ paths. Now we will do the same with plots.

You may remember the integral of a function over an interval: it represents the exact area between the curve and the *x*-axis over the interval. Let's see how to visualize this.

The fillbetween library provides ways to fill areas between plots and axes. You can load it this way:

```
\usepgfplotslibrary{fillbetween}
```

Let's look at the axes and how we can access them as TikZ paths. pgfplots has its own coordinate system that can be accessed using the axis cs prefix. Using this, the plot coordinate system coordinates are translated to TikZ coordinates. So, in TikZ, we can work with a coordinate (axis cs:1,2) which is the coordinate (1,2) in the plot coordinate system, no matter what its TikZ size is.

In the following example, we give a plot a path name. Then, we define a TikZ path with axis cs coordinates, which match the axis. Finally, we use the fill between operation together with \addplot as follows:

```
\begin{axis}[axis lines = center,
  axis equal image, domain = -1.5:1.5]
  \addplot[name path=quartic] {(x^2-1)^2};
  \path[name path=xaxis] (axis cs:-1.6,0)
    -- (axis cs:1.6,0);
  \addplot[darkgray, opacity=0.5]
    fill between[of=quartic and xaxis];
\end{axis}
```

This gives us the following image:

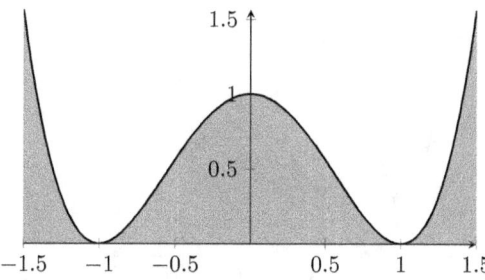

Figure 13.5 – Filling the area below a plot

The whole area between the plot and the axis has been filled. We can limit it to a particular segment by using the `soft clip` option:

```
\addplot[darkgray, opacity=0.5]
    fill between[of=quartic and xaxis,
    soft clip = {domain=-0.5:0.5} ];
```

The modified result is this:

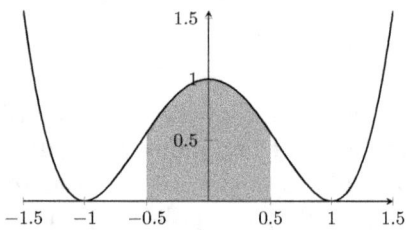

Figure 13.6 – Filling a segment below a plot

We can fill areas between arbitrary curves. In the following example, we fill the area between a cubic curve and a quartic curve that we already know:

```
\begin{axis}[axis lines = center, axis equal,
    domain = -1.5:1]
  \addplot[name path=cubic]    {x^3/5 - x};
  \addplot[name path=quartic]  {(x^2-1)^2};
  \addplot fill between[of=cubic and quartic, split,
    every segment/.style       = {transparent},
    every segment no 1/.style = {gray, opaque}];
\end{axis}
```

This code gives us the following image:

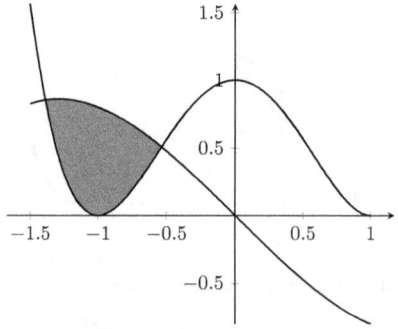

Figure 13.7 – Filling the area between plots

The `split` option is used to split the filling area into several parts, called a **segment**, when the curves have several intersections.

We defined that every segment shall be transparent except segment 1, which shall have a gray color.

In the next section, we will see how to address the intersection points between curves.

Calculating plot intersections

In *Chapter 10, Calculating with Coordinates and Paths*, we calculated the intersection points of TikZ paths. Similarly, we can let `pgfplots` determine the intersections of plots. If you use the `fillbetween` library as we did in the previous section, `pgfplots` will automatically load the TikZ `intersections` library. Otherwise, you can load it yourself.

First, we need to give each plot path a name. Then, we can calculate the intersection points as we did in *Chapter 10*, highlighted here:

```
\begin{axis}[axis lines = center, axis equal,
    domain = -1.5:1]
  \addplot[name path=cubic]    {x^3/5 - x};
  \addplot[name path=quartic]  {(x^2-1)^2};
  \fill[name intersections = {of=cubic and quartic,
    name=p}]
      (p-1) circle (2pt) node [above right] {$p_1$}
      (p-2) circle (2pt) node [left]        {$p_2$};
\end{axis}
```

While we could use `\path`, the `\fill` command is used to produce filled circles at the intersection coordinates.

We get the following plot:

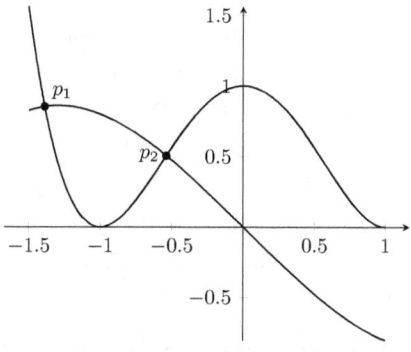

Figure 13.8 – Intersection points of plots

If `name path` doesn't work, such as when the plot path is in a separate scope, you can use `name path global` instead. It should be a unique name to avoid collisions in naming paths.

When we have two functions in as the preceding plot, it's a good idea to add a legend to identify them, so let's do this next.

Adding a legend

When we have several plots or datasets, it can help to identify each plot with a different color and a description. To do this, we can add a **legend**. Typically, this is a box within the plot area containing a symbol or color to identify each plot, along with their descriptions.

To the code of the previous example, we just need to add the following axis option:

```
legend entries = {$\frac{1}{5}x^3-x$, $(x^2-1)^2$}
```

This adds a box with a description of our plot:

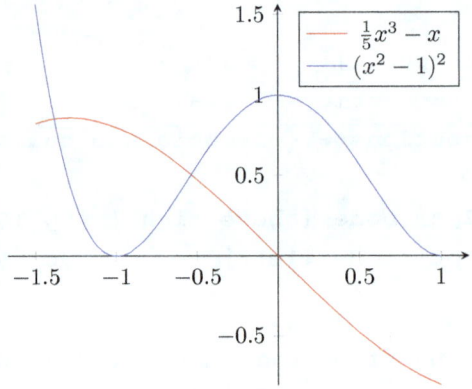

Figure 13.9 – Plots with a legend

To place the legend in the top left of the plot, add `legend pos = north west` to the axis options. Similarly, you can choose `south west` or `south east`, whereas `north east` is the default. If you don't have any whitespace where the legend fits nicely, you can set `legend pos = outer north east`; then, the legend will be placed next to the top-right corner of the plot without overlapping. You can see this in *Figure 14.13*.

The legend entries are horizontally centered by default. Write `legend cell align=left` to align the entries on the left side, as can be seen in *Figure 14.12*, or choose `right` for right alignment, as in *Figure 14.16*.

For more options regarding positioning and appearance, refer to the `pgfplots` manual.

Now, let's examine other options besides Cartesian coordinates.

Using the polar coordinate system

In *Chapter 2*, we talked about polar coordinates. Polar coordinates are just perfect for representing circular or radial symmetric data. Have a quick look back at that chapter, especially at *Figure 2.5*.

To use polar coordinates, we need to load the corresponding library:

```
\usepgfplotslibrary{polar}
```

Then, we have a new `polaraxis` environment. We can use this just like a normal axis, except that the labels, ticks, and grids are now radial. Take a look at this:

```
\begin{polaraxis}
  \addplot[domain=0:180, samples=100, thick] {sin(3*x)};
\end{polaraxis}
```

While a polar plot of sin(x) would give us a simple circle, this relatively simple plot command provides us with the following plot with three leaves:

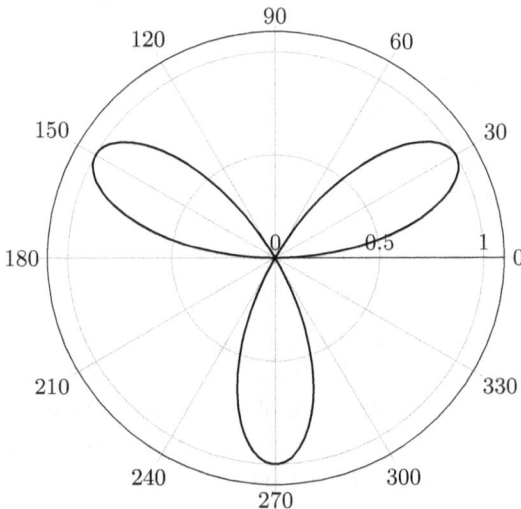

Figure 13.10 – A trigonometric function in a polar coordinate system

We can drive it on and use a large domain, especially with a fractional argument of the sine function, like this:

```
\addplot[domain=0:2880, samples=800, thick] {sin(9*x/8)};
```

This multiply runs around the origin and results in the following image:

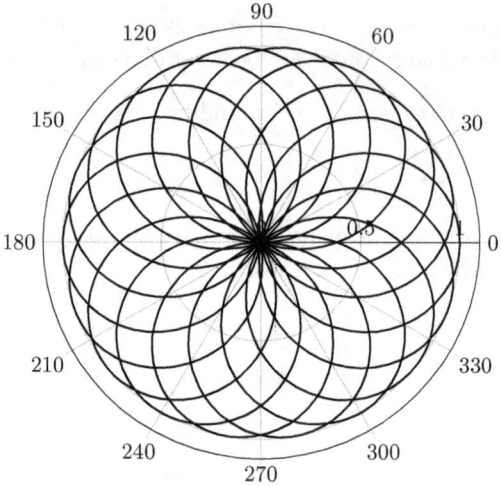

Figure 13.11 – A trigonometric function over multiple times 360 degrees

In summary, polar-defined functions can be very easily defined using the angle and radius. Let's make another polar plot with parametrization in the next section.

Parametric plotting

In *Chapter 10*, we used the `calc` package to draw Archimedean spirals in *Figure 10.8* and *Figure 10.9*. The syntax gets easier with a plotting package, and we get a coordinate system with axes on top.

Instead of using degrees for angles, we can use **radians**. These are an alternative means of angle measurement, happily used especially by mathematicians. Though radian values are simple numbers, we usually express them in multiples of π. For example, a right angle, 90 degrees, would be written as π/2, and 180 degrees are equal to π. We could say 180 degrees is about 3.14 in radians, but we use π. In the same way, 360 degrees equal 2π, and 1,080 degrees is 6π.

We will use radian values and labels in our next plot. For this, we switch the plotting format to radian using the following command:

```
\pgfplotsset{trig format plots=rad}
```

Now, we can use radian values for the domain of the plot, which is calculated with radian arguments. Let's plot an Archimedean spiral, similar to what we did in *Chapter 10* in the *Calculating with coordinates* section.

From a basic math book, we can get the parametrization of a circle. It is x(t)=r*cos(t) and y(t)=r*sin(t), with r as the radius of the circle and t as the angle between 0 and 360 degrees. That's a constant radius, which means a constant distance from the origin. A spiral has an increasing or decreasing distance from the origin as the angle gets higher or lower. In our case, we will print an Archimedean spiral where the distance to the origin is precisely equal to the angle. This is similar to the parametrization of a circle as covered previously: (x,y) = (t*cos(t),t*sin(t)).

In the following plot, we will use a radian domain with radian expressions in multiples of π for tick values and labels. Furthermore, we will rename the variable x to t to avoid confusing the angle with a Cartesian coordinate x value.

Our code is as follows; new things are highlighted:

```
\begin{axis}[axis lines = middle, axis equal,
    domain = 0:6*pi, ymin=-18, ymax=18,
    xtick = {-4*pi,-2*pi,pi,3*pi,5*pi},
    ytick = {pi, 2*pi, 3*pi, 4*pi, 5*pi},
    xticklabels = {$-4\pi$, $-2\pi$,
      $\vphantom{1}\pi$, $3\pi$, $5\pi$},
    yticklabels = {$\vphantom{1}\pi$, $2\pi$,
      $3\pi$, $4\pi$, $5\pi$}
  ]
  \addplot[samples=120, smooth, thick, variable=t]
    ( {t*cos(t)}, {t*sin(t)} );
\end{axis}
```

Compiling this, you get the following image:

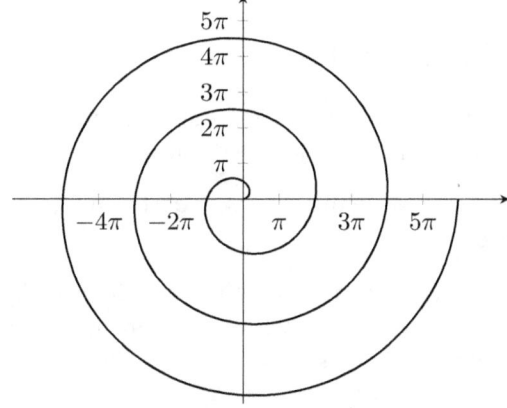

Figure 13.12 – Archimedean spiral

Remember the polar coordinates in the previous section? Let's try it out. A polar plot of an Archimedean spiral is much easier to do like this, this time using degrees and with more loops around the origin:

```
\addplot[domain=0:2880, samples=200, smooth, thick] {x};
```

So, simply having the radius (y) equal to the angle (x) gives us the following polar plot:

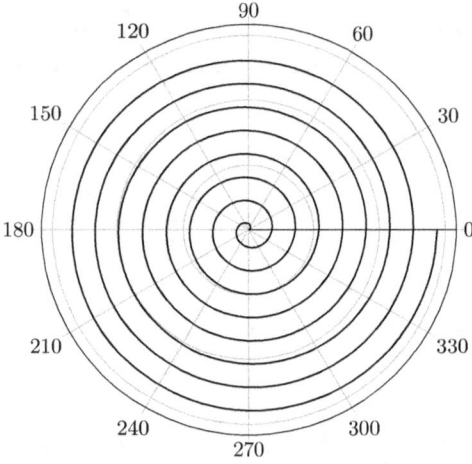

Figure 13.13 – Archimedean spiral plotted in a polar coordinate system

The same can be done with radian angles by setting `data cs=polarrad`:

```
\addplot[domain=0:16*pi, samples=400, smooth, thick,
    data cs=polarrad] {x};
```

Now that we know about parametric plots and radian angles, let's use them in three dimensions.

Plotting in three dimensions

`pgfplots` has impressive 3D plotting capabilities. There are so many customization options that we will leave most of the details to the manual and just go through a few examples here.

We will use the `\addplot3` command similarly to `\addplot`; now, we have functions such as z = f(x,y) or parametrization in x, y, and z.

`pgfplots` easily provides 3D axes, drawn as a box with ticks at the edges by default. One interesting feature is **color maps**: we can improve our 3D visualizations by mapping the z value to a color. To get started, let's load the corresponding library first:

```
\usepgfplotslibrary{colormaps}
```

We continue to use the radian format:

```
\pgfplotsset{trig format plots=rad}
```

We will use a black-and-white color map, where the lowest z values are black, and the higher the z value, the lighter the color. The highest z value will be printed in white.

A typical visualization is a surface plot that draws a mesh representing the function z, which depends on x and y.

This example shows how we can do it:

```
\begin{axis}[
    domain     = -4:4, samples y = 80,
    y domain   = -4:4, samples   = 80,
    colormap/blackwhite, grid ]
  \addplot3[surf] { cos(sqrt(x^2+y^2)) };
\end{axis}
```

There is a little more new syntax to learn here. surf stands for surface plot, and colormap/blackwhite gives us a color mapping so that lower points are displayed in a darker color, which improves the 3D experience. This is a plot of the cosine of the distance to the origin, calculated by sqrt(x^2+y^2); remember Pythagoras. So, basically, it is a cosine rotated around the z-axis, which when plotted gives us the following:

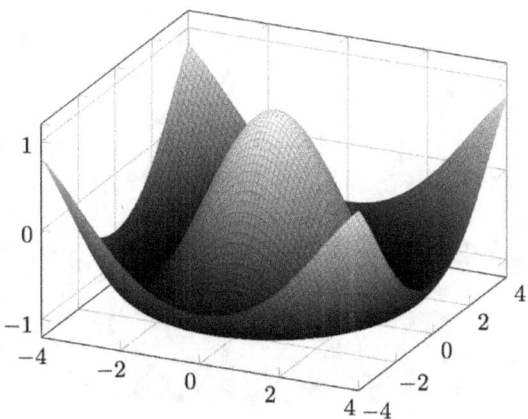

Figure 13.14 – A plot in 3D coordinates

At the plot's edges, we can see that a xy rectangle is used for sampling the coordinates. Suppose we are interested in sampling over a circular area. In that case, we can use a parametrization of radius r and angle t also in 3D, similar to what we did in our spiral plot.

Let's rotate our quartic function from *Figure 13.4* around the *z*-axis. We will have (x,y,z) parametrized, x and y like a circle as mentioned in our 2D *Parametric plots* section, and z will be our quartic function, applied to the radius r. We will use a different, brighter color map, rename the variables again, and omit the axis. We need to use the `z buffer=sort` option, which sorts the points by the value of the z coordinate, ensuring that those points closer to us are rendered after points that are farther away from our viewpoint. We do this so that the front of the plot facing us is not obscured by background points just because of the order of printing. The new syntax elements are highlighted here:

```
\begin{axis}[hide axis, colormap/hot2]
  \addplot3 [surf, z buffer=sort, trig format plots=rad,
    samples=65, domain=-pi:pi, y domain=0:1.25,
    variable=t, variable y=r]
    ({r*sin(t)}, {r*cos(t)}, {(r^2-1)^2});
\end{axis}
```

Mainly thanks to the splendid color map, we get a magnificent surface plot:

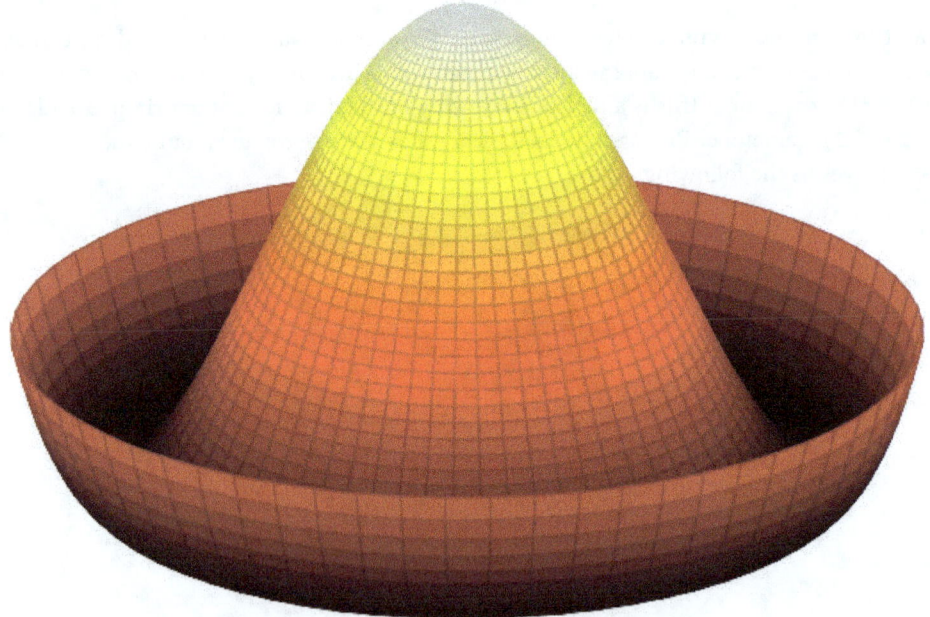

Figure 13.15 – A sombrero plot

This section is just a quick introduction to 3D plotting with `pgfplots` to give an impression of what you can achieve. The `pgfplots` manual is an excellent reference for all such plots with a lot of examples.

Summary

In this chapter, you gained knowledge and skills to visualize data points and mathematical functions in scientific or technical presentations. You can now plot datasets and functions in both two and three dimensions with Cartesian or polar coordinates to present your data in an informative and pleasing way.

The next chapter will teach you to present data using diagrams and charts.

Further reading

The TikZ manual at `https://texdoc.org/pkg/tikz` explains basic plotting in *Part III, Section 22, Plots of Functions*. You can read it online at `https://tikz.dev/tikz-plots`.

Part VI, Data Visualization, covers data point and function plots in depth on more than a hundred pages. This is a new and promising concept; you may consider it as a fresh alternative to `pgfplots`. The quick link is `https://tikz.dev/dv`.

The `pgfplots` package is comprehensively documented in its detailed manual. This excellent reference document contains numerous examples and even tutorials. You can open it at the command line with `texdoc pgfplots` or visit `https://texdoc.org/pkg/pgfplots`.

Did you like the `tikz.dev` deep links to particular sections in past chapters? Then you may appreciate this: while I was writing this book, the `pgfplots` manual was also made available as an online HTML version. You can browse it at `https://tikz.dev/pgfplots`.

The *LaTeX Cookbook* from Packt Publishing, written by me, contains several plot examples in *Chapter 10, Advanced Mathematics*, which motivated me to continue this work here. You will find fascinating 2D and 3D axis styles there. Take a look at `https://latex-cookbook.net/tag/pgfplots` or `https://latex-cookbook.net/chapter10`.

While you can also find plots in the TikZ galleries you already know, such as `https://tikz.net` and `https://texample.net`, there's a whole site dedicated to that excellent plotting package: visit `https://pgfplots.net`.

In addition, there's a gallery of all examples extracted from the manual. You can find it at `https://pgfplots.sourceforge.net/gallery.html`.

`tikz-3dplot` is another package extending TikZ's 3D capabilities. Its author, Jeff Hein, wrote an introduction at `https://latex.net/tikz-3dplot`, and you can read the manual at `https://texdoc.org/pkg/tikz-3dplot`.

14
Drawing Diagrams

As well as standard documents, presentation slides and conference posters often utilize diagrams to represent information visually. This chapter shows you how to create various kinds of diagrams with TikZ.

In this chapter, we'll be covering the following topics:

- Creating flowcharts
- Building relationship diagrams
- Writing descriptive diagrams
- Producing quantitative diagrams

As you already have the tools to create diagrams manually using nodes, styles, positioning, and arrows, this chapter focuses on packages that generate whole diagrams.

Once you have mastered this chapter, you can create colorful and visually stunning diagrams in any context.

Technical requirements

You can find the code examples for this chapter at `https://tikz.org/chapter-14`. On GitHub, you can download them at `https://github.com/PacktPublishing/LaTeX-graphics-with-TikZ/tree/main/14-diagrams`.

We will use the following packages: `smartdiagram`, `sansmath`, `pgf-pie`, `wheelchart`, and `fontawesome5`.

Creating flowcharts

If we want to illustrate a process or a workflow, we can create a **flowchart**. Such a diagram consists of nodes that represent, for example, process steps or decision points and arrows that indicate the process flow. We produced our first flowchart in *Chapter 4, Drawing Edges and Arrows*. Our result was *Figure 4.6*.

This section will look into a handy package that provides quick ways to create flowcharts and other diagrams easily. That package is called `smartdiagram`, and it truly deserves this name. Just look at how fast we can create a flowchart with just a few lines of code now.

First, we have to load the package:

```
\usepackage{smartdiagram}
```

It comes with a `\smartdiagramset` command that is used to customize the diagrams, and it works similarly to `\tikzset` and `\pgfplotsset`, except it is just for smart diagrams. For example, I strongly prefer sans-serif text in diagram nodes, so I can use the following command to get a sans-serif font in a diagram:

```
\smartdiagramset{font=\sffamily}
```

We can also now use a single command to create a flowchart that way:

```
\smartdiagram[diagram type]{comma-separated item list}
```

In the following sections, we will create linear and circular flowcharts, or, as `smartdiagram` calls them, **flow diagrams**.

Linear flow diagrams

The following code prints a horizontal flowchart that illustrates the steps we took in *Chapter 4* to produce a flowchart manually:

```
\smartdiagram[flow diagram:horizontal]{
  Define styles, Position nodes, Add arrows,
  Add labels, Review and refine}
```

This gives us the following diagram:

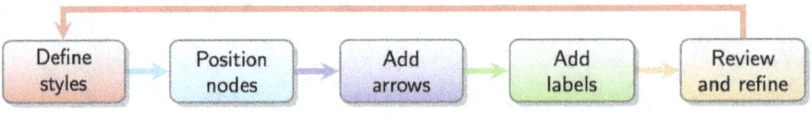

Figure 14.1 – A flowchart depicting a diagram creation process

Note that we don't use a `tikzpicture` environment here; this is implicitly used.

This is super-easy, even for beginners! Node positions, sizes, and colors are predefined, and arrows are as well. You can remove the default long backward arrow by writing `\smartdiagramset{back arrow disabled}` in your preamble or before the diagram. If you keep it, you can change the distance to the nodes, which is 0.5 by default, to some other value, using `\smartdiagramset{back arrow distance=1}`, for example.

The node colors are chosen from a list of 10 default colors: red, cyan, blue, green, orange, yellow, magenta, brown, violet, and teal (in that order), with an opacity amount of 40%. So, the first node color is `red!40`.

More precisely, that's the bottom color, while the top color is white. This gives a top-down gradient color to improve the shading effect, which is also done by default.

Section 8, *Implementation*, of the `smartdiagram` manual, contains the package's source code. It gives some ideas for modifying default `smartdiagram` styles. For example, a `smartdiagram` node is called a **module**. In the code, we find this setting:

```
\tikzset{module/.style={...,
  top color=white, bottom color=\col, ...}}
```

`\col` is the current color from the color list. By appending to the style, we can override settings, such as this:

```
\tikzset{module/.append style = {top color=\col}}
```

Voila! The top and bottom colors are now the same, resulting in a uniform solid color of the nodes. By using TikZ commands again, we can disable the shadows of the nodes by setting the filling to none and scaling them to zero so that they don't even reserve space:

```
\tikzset{every shadow/.style = {fill=none} }
```

As usual, we can combine such style commands in a single `\tikzset` or `\smartdiagramset` command, respectively. Let's do this now with a vertical version of our flowchart. We can add further settings, such as widening text and scaling the shadow to zero, to be sure it's not even taking up space:

```
\smartdiagramset{font=\sffamily,
   text width = 3cm, back arrow disabled}
\tikzset{module/.append style = {top color=\col},
   every shadow/.style = {fill=none, shadow scale=0}}
\smartdiagram[flow diagram]{
  Define styles, Position nodes, Add arrows,
  Add labels, Review and refine}
```

Note that we omitted the `horizontal` keyword, so by default, we get a vertical flowchart, as follows:

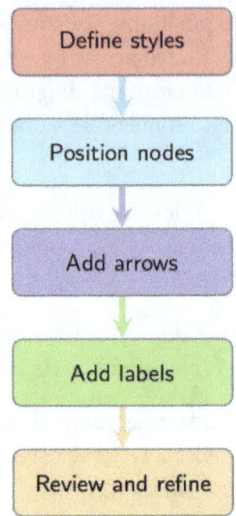

Figure 14.2 – A vertical flowchart with flat colors and without shadows

For smart diagrams, there are many customizing options available. The manual explains all of them; I decided to explain most here so that you have enough tools at hand in this chapter.

We will see many options with customized values in the upcoming examples so that you can understand how to apply them.

The following options are to modify colors:

- `set color list` is a comma-separated list of colors that redefines the list of colors for the nodes. We will use this for *Figure 14.3*.
- `uniform color list` is to choose a single color for all nodes. We have to specify the number of nodes; refer to the code for *Figure 14.7* to see how to do this.
- `use predefined color list` switches back to the original default color list.

By default, arrows have the same color as the node they point to. You can change this as follows:

- `uniform arrow color` decides to have a single color for all arrows if set to `true`. It is `false` by default.
- `arrow color` is the color to be used when `uniform arrow color` has been set to `true`.

There are more options to modify arrows:

- `arrow tip` can be used to choose the arrow tip, which is `stealth` by default. See *Chapter 4, Drawing Edges and Arrows*, for arrow tips.

- `arrow style` can be used to define a new style for the arrows.
- `arrow line width` stands for the thickness of the arrow line, which is 1 mm by default.

`smartdiagram` also provides diagram type-specific options. The following options are supported for linear and circular flow diagrams. These options are to customize the nodes, which are called modules here:

- `module minimum width` is the width of a module, initially 2 cm.
- `module minimum height` is the height of a module, initially 1 cm.
- `module x sep` is the horizontal distance between modules, which is 2.75 by default.
- `module y sep` is the vertical distance between modules, which is 1.65 by default.
- `module shape` is the shape. You can choose any shape you saw in *Chapter 3, Drawing and Positioning Nodes*. If the shape comes from a particular TikZ library, you need to load that. By default, the module shape is a rectangle with rounded corners.

Then, we have some text and color options:

- `font` is a font command for the module text, initially `\small`, but you can use further commands and combine them, such as `\sffamily\Large`.
- `text width` sets the text width, which is 1.75 cm by default.
- `text color` chooses the color of the text. Initially, it's black, of course.
- `border color` sets the border color, which is gray by default.

There's another flowchart type called **sequence diagram**. In this, the nodes are the arrows, having an arrow shape. The following example creates such a sequence. Particular customization options are highlighted:

```
\smartdiagramset{
   sequence item font size = \sffamily\Large\strut,
   set color list = {red!80, red!60, red!45, red!30} }
\tikzset{module/.append style = {top color=\col} }
\smartdiagram[sequence diagram]{
   Styles, Positions, Arrows, Labels}
```

What catches the eye the most is `\sffamily\Large\strut` as `font size`. However, we can use any `font` command here. I intentionally inserted the `\strut` command, an invisible vertical line that matches the text height, including the depth of the descenders. That's an easy way to align the node texts vertically; we need it because the first node's text contains *y*, affecting the baseline. Look closely at the node text alignment in *Figure 14.1*; then, you will see what I mean – the baseline alignment was slightly off there. Try it without!

We also applied further options in this example, as promised. In particular, we chose different node colors and disabled the color gradient, as explained previously.

The output of this example is the following:

Figure 14.3 – A sequence diagram with custom colors

For a sequence diagram, there are specific options to customize the sequence items. By changing the following options, you can adjust the appearance of the sequence nodes:

- `sequence item width` is the minimum width of an item node; it's 2 cm by default
- `sequence item height` is the minimum height of an item; it's 1 cm by default
- `sequence item border color` is initially gray, and you can modify it to change the color of the borderline
- `sequence item border size` is the width of the borderline
- `sequence item fill opacity` is initially 1, so we won't have transparency, but you can change that, which would make sense if we have overlapping in the diagram
- `uniform sequence color` can be set to `true`; then, you will get a single color for all items
- `sequence item uniform color` can be used to choose a uniform color for all items

These options can adjust the text in the item nodes:

- `sequence item font size` can be used to change the text font; we used it for *Figure 14.3*.
- `sequence item text width` sets the text width; it's 1.9 cm by default.
- `sequence item text opacity` can be used to make the text transparent. Initially, it has the value 1, which means it's opaque without any transparency.

There are many options. However, the more options we have, the more knobs we can turn to achieve an optimally adapted result without resorting to additional TikZ commands.

Now, let's look at circular diagrams.

Circular flow diagrams

For a flow diagram with an arrow pointing back to the first node, a **circular diagram** would be a natural choice. We will take the command for *Figure 14.1* and only change the diagram type to `circular diagram` and `clockwise`:

```
\smartdiagram[circular diagram:clockwise]{
  Define styles, Position nodes, Add arrows,
  Add labels, Review and refine}
```

This is what we get from it:

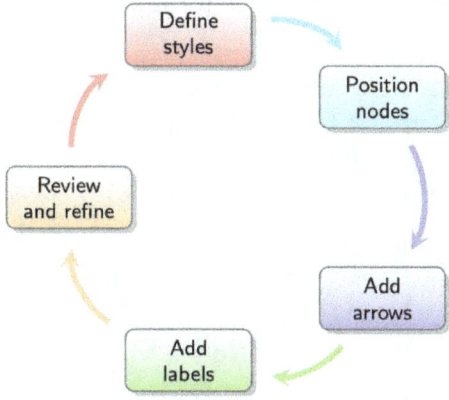

Figure 14.4 – A circular diagram

Again, we quickly got a nice diagram with little work.

In addition to the customization options of flow diagrams, we have the `circular distance` option, which is the radius of the circle where the nodes are placed and has a value of 2.75 cm by default.

Similar to the linear flow diagrams, if we don't want to have the arrow back to the first node, we can disable it by setting `circular final arrow disabled` to `true`.

When we omit the `clockwise` keyword, the diagram will run counterclockwise.

While we have many ways to customize our diagrams, they are still a single flow. We can add elements and annotations. Please take a look at the *Additions* section in the `smartdiagram` manual, which explains such modifications in detail.

We will continue with circular-looking diagrams in the following section, which are not process flows anymore, but illustrations of connections or associations of objects.

Building relationship diagrams

In *Chapter 6, Drawing Trees and Graphs*, we encountered a special relationship diagram, the mind map. The `smartdiagram` package offers other diagram types that indicate connections between concepts or objects. Such diagrams are naturally not linear – for example, a mind map can be very complex, such as in *Figure 6.15*.

Since they are focused and aesthetically pleasing, we will focus on **circular relationship diagrams**, which have a central concept and related concepts placed around them.

Our first diagram type shall be a **bubble diagram**. It's like a mind map – there is a central concept or object, and related concepts are placed around it in a circular manner and shape. The following code illustrates it:

```
\smartdiagramset{bubble node font=\sffamily\LARGE,
  bubble center node font=\sffamily\Huge}
\smartdiagram[bubble diagram]{Diagrams,
    Nodes, Edges, Arrows, Labels, Colors}
```

This code generates the following diagram:

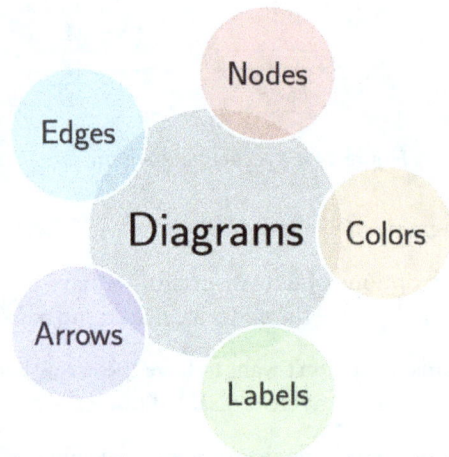

Figure 14.5 – A bubble diagram

The diagram nodes are made from a comma-separated list of items, where the first item is the text for the center node, and the other items go to the outer bubbles in counterclockwise order.

We will set the font with different options from the font options in the other diagram types. Each diagram type has its own option names.

We have these options for the diagram's center:

- `bubble center node font` contains font commands for the center node; we used it in *Figure 14.5*. Initially, it's `\large`.
- `bubble center node size` is the center node size, which is 4 cm by default.
- `bubble center node color` is for the color of the center circle. It is `lightgray!60` by default.
- `distance text center bubble` is the distance from the center text to the center node border, initially 0.5 cm. If you increase or decrease that value, the center bubble gets larger or smaller.
- `distance center/other bubbles` is the distance between the centers of the central bubble and the other bubbles, which is 0.5 cm by default.

The outer bubbles can be customized by redefining the following options:

- `bubble node font` is the font choice for the bubble node texts.
- `bubble text opacity` stands for the opacity of the node texts, which is 0.8 by default.
- `bubble node size` is the minimum size of the outer bubbles. If a lot of text is used, the bubble can get bigger. It is 2.5 cm by default.
- `bubble fill opacity` is initially 0.5 and stands for the opacity; here, you can tweak how transparent the bubbles appear.

Other circular diagrams are the so-called **constellation diagrams**. In our case, these will be diagrams of **satellite** nodes placed around a **planet** node, like in orbit. A **connected constellation diagram** places all satellite nodes in orbit, connected by curved lines.

The following short code snippet produces such a diagram and shows how we can apply some custom settings:

```
\smartdiagramset{planet font=\sffamily\LARGE,
  planet text width=2.2cm,
  satellite font=\sffamily}
\smartdiagram[connected constellation diagram]{
  Drawing diagrams, Define styles,
  Position nodes, Add arrows, Add labels}
```

This generates the following diagram with a large center node, the planet, and four satellite nodes around it:

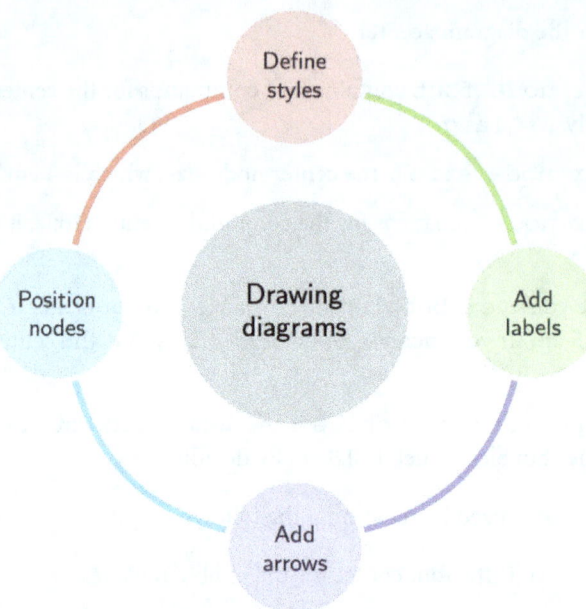

Figure 14.6 – A connected constellation diagram

Again, the first item is the text for the center node, the planet, and the order of satellites is counterclockwise.

The following options can be used for the planet:

- `planet font` is for the font size and shape of the planet's text; initially, it's `\large`.
- `planet size` is the planet's minimum size, initially 2.5 cm. If the planet node text gets too big for it, the node will be enlarged.
- `planet color` is the planet's color; if you don't change that, it will be `lightgray!60`.
- `planet text width` is for the width of the text; it's 1.75 cm by default.
- `distance planet-text` is the distance between the planet's text and its border, initially 0.5 cm.
- `distance planet-connection` is the distance between the planet's border and the arrows, initially 1 mm.

There are similar options for the satellite nodes:

- `satellite font` is for the font size and shape of the text in satellites.
- `satellite size` is the minimum size of the satellites, initially 1.75 cm.
- `satellite text width` is for the width of the text in them; it's 1.5 cm by default.

- `satellite text opacity` is for the opacity of their node texts in them. Initially, it's `0.8`, meaning it's a bit transparent and appears lighter than the planet text.
- `satellite fill opacity` is the opacity of the satellites. It is `0.5` by default, meaning it's 50% transparent. That may be useful when satellites overlap. Note that this way, colors appear lighter than without transparency.
- `distance planet-satellite` is the distance between the planet and the satellites; it's 3.75 cm by default.

You see, we have many ways to tweak diagrams.

The constellation diagrams use the same color list as the other smart diagrams; we can modify this in the same way as we did for *Figure 14.3*. In addition, we can define the colors and line widths of the connections as follows:

- `uniform connection color` can be set to `true`; then, all connection lines will have the same color
- `connection color` can then be used to define the color of the connections
- `connection line width` is the line width of the connections; it's 1 mm by default

Without the `connected` keyword, a constellation diagram will have arrows from the planet to each satellite instead of connection lines between satellites. The preceding connection color and line options are then applied to the arrows.

Let's create such a diagram. For the exercise, we will change all nodes' shapes and colors. Let's go with octagon shapes. At first, we load the corresponding shapes library:

```
\usetikzlibrary{shapes.geometric}
```

Now, we can use `\tikzset` to modify planet and satellite styles. Specifically, we choose polygons with eight sides as the node shape and reduce the `inner sep` value, which is the distance between the node text and border:

```
\tikzset{ planet/.append style={regular polygon,
  regular polygon sides=8, inner sep=6pt},
  satellite/.append style={regular polygon,
  regular polygon sides=8, inner sep=0pt} }
```

Then, we apply specific `smartdiagram` settings, such as a huge planet font, all in sans-serif, and 40% opacity green and red colors:

```
\smartdiagramset{planet font=\sffamily\Huge,
  planet color=green!40,
  satellite font=\sffamily,
  uniform color list = red!40 for 8 items}
```

Now, everything is prepared, and we can create the diagram, again with the first list entry as the planet text:

```
\smartdiagram[constellation diagram]{TikZ,
  pgfplots, smartdiagram, hobby, tikzducks,
  tikzlings, tikzpeople, tikzmark, tikz-ext}
```

With this preparation, we get the following diagram:

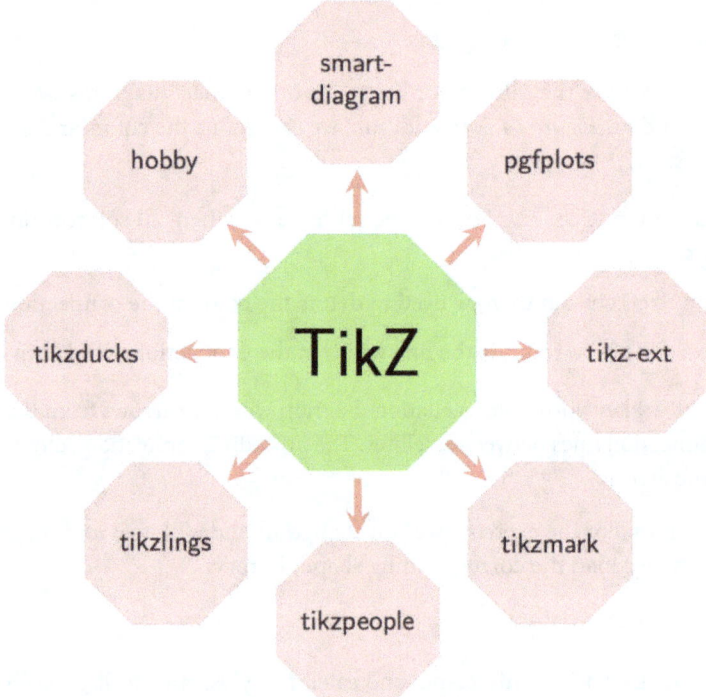

Figure 14.7 – A constellation diagram with arrows

For each satellite, we get an arrow from the planet to that satellite. It has the same color as the corresponding satellite. The arrow distances can be tweaked like this:

- `distance planet-connection` is initially 1 mm and is the distance between the planet border and the arrows.
- `distance satellite-connection` is the distance between the border of satellites and the corresponding arrow; initially, it's 0.75 mm.

Now that we have learned a lot about arranging concepts, let's turn to diagrams that explain concepts.

Writing descriptive diagrams

A **descriptive diagram** typically represents terms and explanations and connects or aligns them for illustration.

You know the standard LaTeX description environment. Let's create a diagram in a similar layout to make it visually stunning.

First, we choose the fonts, a very large font for the items and a small font for each description, all in sans-serif font:

```
\smartdiagramset{description title font=\sffamily\LARGE,
  description font=\sffamily\footnotesize}
```

Now, we use the descriptive diagram type of `smartdiagram` and give a list of pairs of titles and descriptions, each pair in braces. Remember, since the comma is used to separate list items, we have to use additional braces if a description contains a comma itself. Also, end the list with a comma, since `smartdiagram` uses it to properly parse all items. We can come up with a code to describe PGF and TikZ:

```
\smartdiagram[descriptive diagram]{
  {PGF, {Portable Graphics Format, package for
         creating graphics in \LaTeX{} documents}},
  {TikZ, {User-friendly frontend for PGF}},}
```

That already gives us a nice diagram, fancy enough to be on a presentation slide:

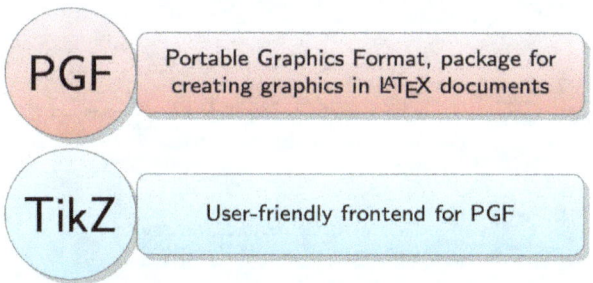

Figure 14.8 – A descriptive diagram

The look reminds us of a **bullet list**. Indeed, we can get more creative and use it like an `itemize` list with customized bullets. To do this, we can load the incredible `fontawesome5` package, which allows us to use version 5 of the famous **Font Awesome** icon library:

```
\usepackage{fontawesome5}
```

Now, we have access to hundreds of icons to add to our drawings. Browse the icon list in the `fontawesome5` manual to find the icon and its corresponding LaTeX command. I did this and came up with the following symbols for a bullet list that explains tasks for drawing diagrams:

```
\smartdiagram[descriptive diagram]{
  {\faLightbulb[regular],{Identify purpose and message,
    gather information and data}},
  {\faProjectDiagram, {Select diagram type,
    define node shapes, colors, and text styles}},
  {\faPencil*, {Draw nodes, insert text, draw arrows,
    add labels}},
  {\faAlignLeft, {Align nodes, refine positioning}},
  {\faRedo, {Fine-tune, review and revise}}, }
```

This results in the following diagram:

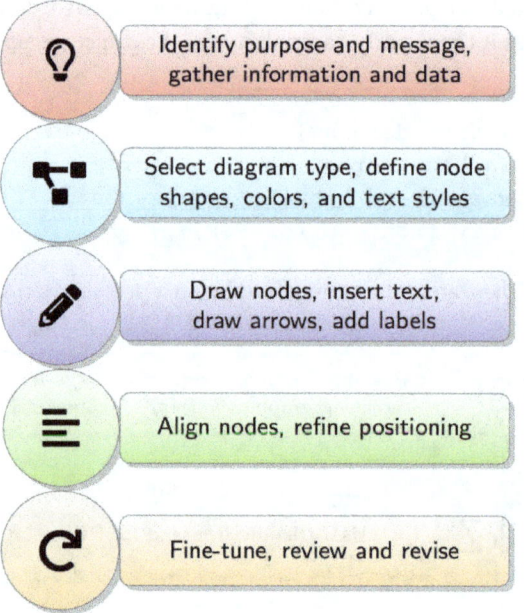

Figure 14.9 – A customized bullet list diagram

Since version 6 of Font Awesome has already been released, we may expect a LaTeX package soon.

The last two diagrams did not imply a particular order. With a **priority descriptive diagram**, we can indicate an order, priority, or dependency. Let's demonstrate this with what we started in *Figure 14.8* by showing that TikZ is built on PGF and adding more dependencies we already know.

We can use the same options for the descriptive diagram, so let's first select the font and width of the description nodes:

```
\smartdiagramset{description font=\sffamily\Large,
  description text width = 1.9cm,
  description width = 2cm}
```

Now, we add the diagram, listing the TikZ-related concepts we know, and order them from the lower layer to the upper layer:

```
\smartdiagram[priority descriptive diagram]{
      \TeX, \LaTeX, PGF, TikZ, pgfplots}
```

This diagram looks as follows:

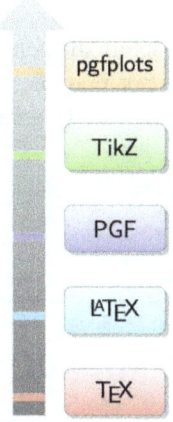

Figure 14.10 – A priority descriptive diagram

Even though PGF, TikZ, and `pgfplots` can be used with `p-reg` and plain `P-reg` too, these are abstraction layers for us as LaTeX users.

In the next section, we will continue with that top layer, using `pgfplots` to create diagrams based on numerical quantities.

Producing quantitative diagrams

Until now, our diagrams presented objects in order or in relation to each other. Now, we will visualize and compare actual values utilizing diagrams.

Line charts

Numerical data is often a series of values measured over time. These values can be displayed as data points on a plane, with the *x* axis for the time and the *y* axis for the values. Connecting lines show the trend over time. This is called a **line chart**. Such a chart can contain multiple line series to display different datasets to see them in relation. The *x* axis doesn't need to represent time; it could be any other base value, such as age, weight, or other data to correlate with values.

In *Figure 12.5*, we saw how a line chart could be plotted, and *Chapter 13*, *Plotting in 2D and 3D*, showed tools to plot in a coordinate system. `pgfplots` provides a perfect setting to display values in a plane by lines or bars with annotations.

This section will compare the graphics packages TikZ, PSTricks, and MetaPost regarding their popularity among users. It's not comparing excellence because all three are excellent; it's just how much we see them used in the field.

Google Trends is an exciting application to explore the popularity of keywords among search engine users over the years. The output is a line chart of a popularity score between 0 and 100. This is what we get when we compare the `tikz`, `pstricks`, and `metapost` keywords:

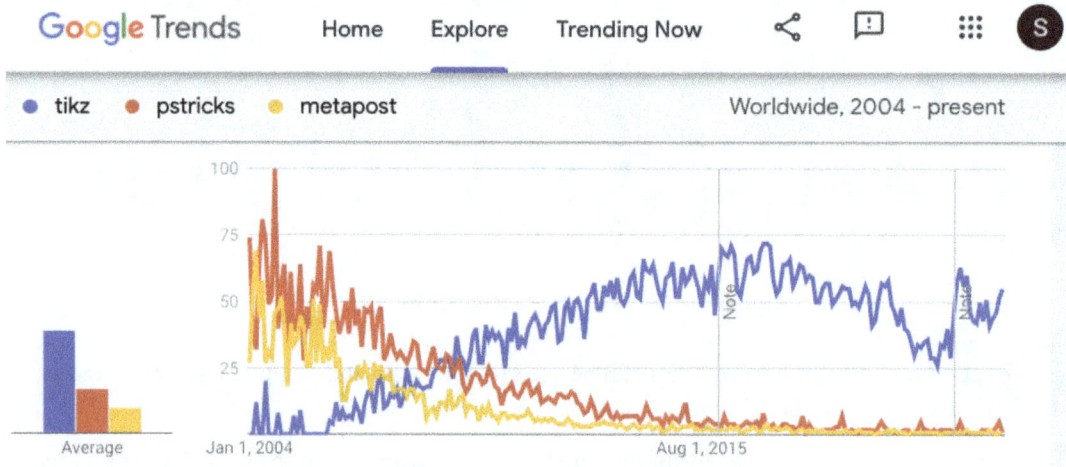

Figure 14.11 – A Google Trends chart for LaTeX graphics packages

You can look at it yourself at `https://trends.google.de/trends/explore?date=all&q=tikz,pstricks,metapost`.

I was baffled the first time I saw how TikZ's popularity has developed over time. Google Trends offers to download the data in CSV format, which stands for **comma-separated values**. I did this, calculated the average values per year, and took these values as samples for the upcoming diagram examples. While the drop on the right clearly looks interesting, representing the COVID-19 pandemic of 2019–2022,

I chose the years when TikZ started gaining popularity, 2007–2013, so we will have only seven data points. That's just for a practical reason to keep our examples short and small.

Let's set up our coordinate system. Remember that I prefer sans-serif diagram fonts, so I set them for all nodes:

```
\tikzset{every node/.style={font=\sffamily}}
```

pgfplots prints the tick labels in math mode, so they will be with serifs and look different. I like to use the sansmath package, which allows us to choose sans-serif fonts in math expressions. This is the way we can apply it to the pgfplots tick labels:

```
\usepackage{sansmath}
\pgfplotsset{tick label style = {font=\sansmath}}
```

By default, pgfplots uses commas as thousands separators, which would look strange in year values, so we will disable it in our axis setting. Finally, let's have a title and a legend in the upper-left corner – in other words, northwest. So, this shall be our axis:

```
\begin{axis}[title  = Keyword popularity in Google trends,
  x tick label style =
    {/pgf/number format/set thousands separator={}},
  legend pos = north west,
  legend cell align=left]
```

Now, we can plot three coordinate series with the values I calculated previously from Google Trends. The first one is for TikZ:

```
\addplot coordinates { (2007,16) (2008,19) (2009,30)
   (2010,36) (2011,42) (2012,48) (2013,55) };
```

Then, we plot the PSTricks values:

```
\addplot coordinates { (2007,39) (2008,28) (2009,24)
   (2010,19) (2011,15) (2012,12) (2013,8) };
```

Finally, we plot the data for MetaPost:

```
\addplot coordinates { (2007,22) (2008,13) (2009,11)
   (2010,8) (2011,6) (2012,4) (2013,4) };
```

We add the legend with the keywords in the same order as we did the plots, and then we close the axis environment:

```
    \legend{tikz, pstricks, metapost}
\end{axis}
```

Remember, as with the previous chapter, this has to be in a `tikzpicture` environment. This results in the following diagram:

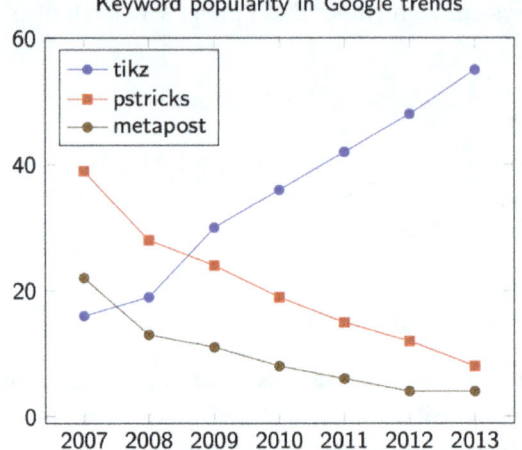

Figure 14.12 – A line chart representing keyword popularity over time

`pgfplots` automatically chooses colors and markers to distinguish the graphs from each other and to display colors and markers accordingly in the legend.

You can use the customization options you learned in the previous chapter or look them up in the manual.

It may be interesting to know whether the graphics' popularity with TeX and LaTeX changed over time. We can change this diagram to a **stacked line chart** to see this. For this, we add the following options to the axis environment:

- `stack plots=y` so that the line graphs are stacked above each other. `stack plots=x` would do the same along the *x* axis, which would make sense when the chart is horizontally oriented, meaning that we focus on *x* values for *y* categories.
- `area style` is for filling the area under a plot.
- `legend pos = outer north east` is to get the legend outside the chart so that it doesn't overlap.

Furthermore, we will add the `\closedcycle` command to each plot to get a closed area for each one that can be filled, so the code goes like this:

```
\addplot coordinates { (2007,16) ... } \closedcycle;
```

This results in the following diagram:

Figure 14.13 – A stacked line chart showing cumulated values and relative proportions

We now can see that the overall popularity stays roughly the same; just the share of each package changes over time.

Apart from data points connected by lines, displaying values by vertical or horizontal bars is very popular. Let's look at this.

Bar charts

If we focus more on comparing categories, we can choose a **bar chart** for visualization. Here, the data is represented by either vertical rectangular bars, where the height stand for the values, or horizontal bars, where the width represent the values. Several categories can be grouped next to each other, and such groups can iterate over a base value such as time.

Let's take the date from the previous section to see how this works.

We will take our line chart example from *Figure 14.12* and modify the axis options in the following way, with our changes highlighted:

```
\begin{axis}[title   = Keyword popularity in
  Google trends,
  ybar, bar width=2mm,
  x tick label style =
    {/pgf/number format/set thousands separator={}},
  legend pos=north west,
  legend cell align=left ]
```

The main point is that we simply added the ybar keyword to get vertical bars in the *y* direction. For horizontal bars, we use xbar. Then, we just reduced the bar width, since we want to have space for three bars next to each other for every year. This time, we positioned the legend in the top-left corner because there was some white space.

Everything else is the same, including the \addplot commands with the coordinate values. This is the bar chart we get:

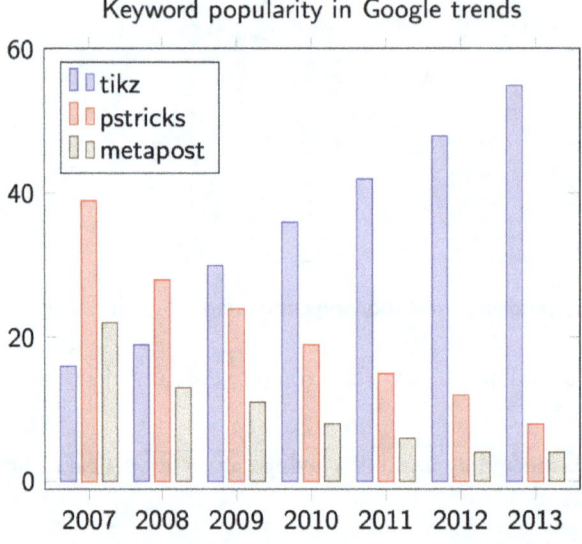

Figure 14.14 – A bar chart visualizing relative and absolute values over time

We can see that the proportions of the keywords change over time, but it's hard to see whether the overall popularity increased or decreased. So, we may be interested in displaying the cumulative values to see how the total graphics package polarity evolved over time. That's the job of a **stacked bar chart**. And it's not complicated at all; just write ybar stacked instead of ybar. To get the legend out of the way, we move it to the top-right corner outside the plot area by changing legend pos to outer north east. The diagram changes as follows:

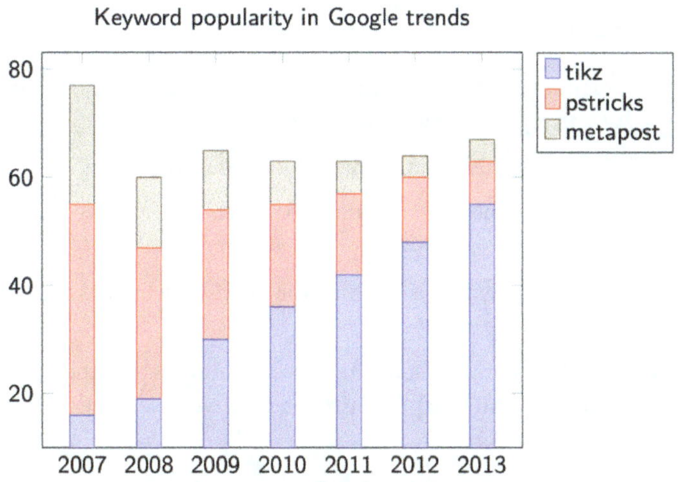

Figure 14.15 – A stacked bar chart showing cumulated values and relative proportions

Now, we can see, like in *Figure 14.13*, that the overall graphics package popularity in the Google search stays roughly the same; it's just that TikZ got a bigger share over time at the expense of the others.

I mentioned horizontal bar charts at the beginning of this section. We get them by writing `xbar` instead of `ybar`, or `xbar stacked` instead of `ybar stacked`. However, the setup of the coordinate axes may need to be different. For example, we can have categories on the left side along the *y* axis, and corresponding values at the *x* axis, so when the values rise, the bars grow to the right side.

In *Chapter 9, Creating Graphics*, of the *LaTeX Cookbook*, I designed a horizontal bar chart to display the number of contributions in forums on `LaTeX.org`. Inspired by this, we will look at an example of creating a nice-looking horizontal bar chart, where we omit axes to focus on the values.

The data will consist of the number of search results for the keywords TikZ, PSTricks, and MetaPost in the `LaTeX.org` forum.

Our axis setup shall be the following:

```
\begin{axis}[title   = Keyword popularity on LaTeX.org,
  height = 6cm, enlarge y limits   = 0.6,
  xbar,
  axis x line = none,
  y axis line style = transparent,
  ytick = data, tickwidth = 0pt,
  symbolic y coords = {TikZ,PSTricks,MetaPost},
  nodes near coords,
  nodes near coords style = {font=\sansmath},
  legend cell align = right ]
```

With the axis options, we achieve the following, in this order:

1. We reduce the axis height to 6 cm because we will have only three *y* values. For the same reason, we shrink the *y* axis by using a scaling factor of 0.6.
2. With `xbar`, we determine that it's a horizontal plot in the *x* direction.
3. We remove the *x* axis line completely, including the labels, because we only want to place the values next to the bars.
4. Instead of also removing the *y* axis, we just hide the *y*-axis line by making it transparent. This way, we keep the *y* labels.
5. We set `ytick` to `data` to generate tick marks at every coordinate of the first plot but nowhere else, as mentioned in the previous chapter. Furthermore, we hide the tick lines by setting their width to zero.
6. We choose symbolic *y* coordinates representing TikZ, PSTricks, and MetaPost.
7. We write the *x* values directly near the bars and choose sans-serif math as the font style.
8. Finally, we align the legend entries on the right-hand side.

Now, we can plot the data, this time using symbolic coordinates defined for the *y* axis. We start with the number of posts per keyword:

```
\addplot coordinates { (2750,TikZ) (1568,PSTricks)
  (69,MetaPost) };
```

Then, we plot the number of topics per keyword:

```
\addplot coordinates { (1197,TikZ) (585,PSTricks)
  (41,MetaPost) };
```

We add the legend and close the `axis` environment:

```
  \legend{Posts,Topics}
\end{axis}
```

As usual, the plotting itself is easy, but we may have to consult the `pgfplots` manual for the axis design to search for suitable options to adjust.

That code generates the following diagram:

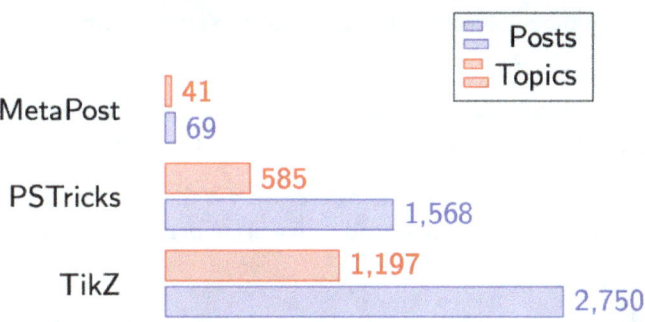

Figure 14.16 – A horizontal bar chart with symbolic coordinates

Also, horizontal bars can be stacked by writing `xbar stacked` instead of just `xbar`. With our settings nodes near `coords`, the values will be placed inside the bars. If we omit MetaPost now, since the bars are too small for it and it's rarely used on `LaTeX.org`, we would get this stacked bar chart:

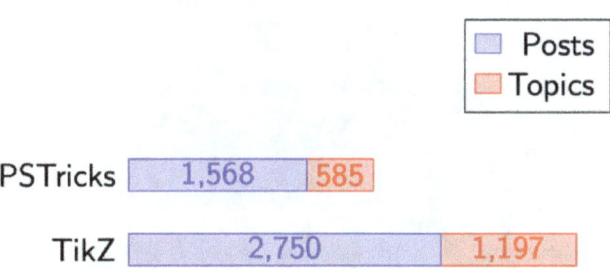

Figure 14.17 – A stacked horizontal bar chart

In a stacked bar chart, a rectangle is divided into segments. In the next section, we will divide a circle to show proportions.

Pie charts

We can also display categories and values in a circular way. If our data permits it, we can use a **pie chart**, where every category is displayed like a slice of a pie. Two points are commonly considered:

- The categories should be disjoint; otherwise, slices would have to overlap
- The values of the categories can be summarized into a total, such as 100%, so that the circle closes

In other words, a pie chart shows the relative size of a category as a part of a whole, like a percentage breakdown.

The `pgf-pie` package can help to easily create such charts. I have explained this in the *LaTeX Cookbook*, so I strongly recommend reading its freely available sample in *Chapter 9*, *Creating Graphics*, at `https://latex-cookbook.net/chapter9`.

However, let's have a quick example here. In honor of the main contributors to the `TikZ.net` gallery, here is a pie chart representing the percentage of contributions they made.

We need to have a `tikzpicture` environment; we will use that also to choose a sans-serif font again. The pie is then created by a single command with a list of percentage/category pairs, as follows:

```
\begin{tikzpicture}[every node/.style={font=\sffamily}]
  \pie{ 42/Izaak Neutelings,
        21/Janosh Riebesell,
        17/Alexandros Tsagkaropoulos,
        10/Efraín Soto Apolinar,
        10/Other authors }
\end{tikzpicture}
```

That produces the following diagram:

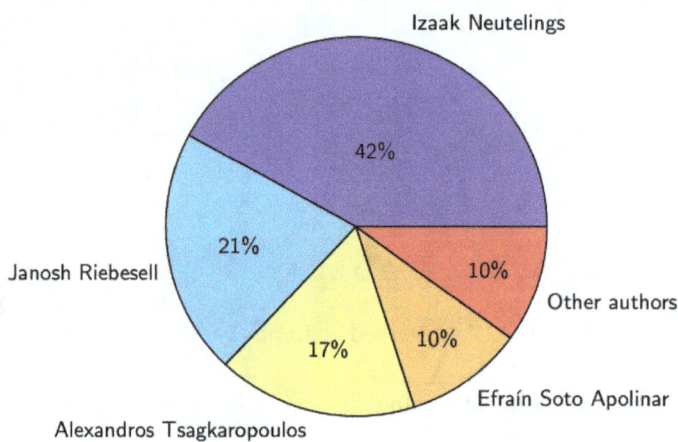

Figure 14.18 – A pie chart

Instead of using percentage values, you can list absolute values. In that case, write `\pie[sum=auto] {...}`; then, the sum is calculated, and the pie slices are displayed, representing their share of the whole data. However, the absolute values are printed then, possibly making it harder to compare them.

The pgf-pie package has many options to customize appearance, colors, positioning, and more. If you want to use it, look at the free *LaTeX Cookbook* chapter or the pgf-pie manual.

The next section covers a similar approach.

Wheel charts

Like a pie chart, a **wheel chart** visualizes data circularly for easy comparison. The wheelchart package written by Matthias Floré provides a single command to draw such charts. The basic syntax is the following:

```
\wheelchart[options]{data}
```

The options can be a list of key=value settings, such as radius, colors, font, and other styling options. For data, the command expects a comma-separated list of items in a shape such as value/style/data, similar to the pgf-pie syntax and the syntax in \foreach loops. However, each item can contain up to 26 data parts because it's alphabetically encoded internally. You should use at least three kinds of data here – a numerical value, color, and corresponding text.

Sounds complicated? Let's look at it with the values from *Figure 14.18* as an example to illustrate the usage. Use the following command in a tikzpicture environment, exactly like the \pie command in the previous section:

```
\wheelchart [middle={{\LARGE TikZ.net}\\contributions},
  inner data = {\scriptsize\WCperc}, inner data sep=0.3,
  wheel lines = white]
    {42/red/Izaak\\Neutelings,
     21/orange/Janosh Riebesell,
     17/yellow/Alexandros\\Tsagkaropoulos,
     10/green/Efraín Soto Apolinar,
     10/blue/Other authors}
```

This command generates the following output:

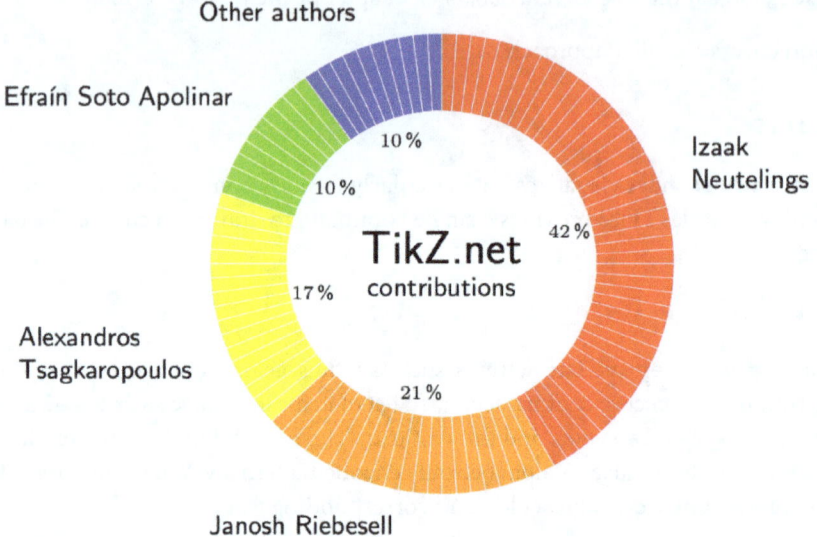

Figure 14.19 – A wheel chart

`middle value` is the text in the middle. `\Wcperc` is the variable containing the current percentage in the loop that we put into the `inner data` key, which is printed on the inner side of the wheel. `inner data sep` is to adjust the distance of `inner data` to the wheel. The other data, the names, are printed outside of the wheel.

Take this example as inspiration; the package can do much more than we can cover here. The manual, available at `https://texdoc.org/pkg/wheelchart`, lists all the available options and shows a lot of examples, including styling as a pie chart, labels with nice-looking connection lines, and stunning slice shapes. Note that you can combine several wheel charts with different radii in a single image to visualize complex data.

Summary

This chapter covered a wide range of diagram types with many examples that can be the foundation for your diagrams.

You learned to quickly create flowcharts, perfect for breaking down complex processes into easy-to-follow steps. With relationship diagrams, you can now visualize the connections and relationships between different concepts. Your descriptive diagrams can make your ideas easier to understand. With quantitative charts, you can showcase your data in the best way to visualize and compare values.

With the diagram tools you studied in this chapter and the TikZ knowledge you acquired while reading this book, you are ready to create beautiful and informative graphic illustrations for your documents.

After all this hard work, you deserve some fun. That shall be the motto for the next chapter!

Further reading

The `smartdiagram` package provides a detailed manual with a complete reference to all settings and even its source code. You can open it on your computer at the command line with `texdoc smartdiagram` or by visiting https://texdoc.org/pkg/smartdiagram.

The *LaTeX Cookbook* by Packt Publishing and written by me has a *Building smart diagrams* section in *Chapter 9, Creating Graphics*. It shows more `smartdiagram` examples, just not as customized as in this chapter, and with detailed explanations and references. You can find the entire section at https://latex-cookbook.net/9-1-building-smart-diagrams/.

The *LaTeX Cookbook* explains in detail how to create a complex flowchart from scratch with just standard TikZ tools. This section is also available online; you can read it at https://latex-cookbook.net/9-2-constructing-a-flowchart. You can also find pie charts and Venn diagrams examples in the same book and on its website.

The `pgf-pie` manual can be found at https://texdoc.org/pkg/pgf-pie.

Also visit https://tikz.net and https://texample.net, which contain dozens of diagram examples of all kinds.

15
Having Fun with TikZ

Congratulations on making it to the final chapter! You worked hard through this book and have become seasoned and proficient in TikZ. This chapter will reward you with entertaining drawings you can try, modify, and combine with your pictures. Mainly, it demonstrates how skilled TikZ users had fun programming add-on packages and sharing them with the TikZ community.

The final chapter of our journey explores the following topics:

- Drawing cute creatures
- Playing and crafting
- Drawing world flags

We will walk through examples to see how to use the packages and understand how to customize the drawings, while you can find the complete reference of all features in the package manuals.

Technical requirements

You can find the drawings with the entire source code at `https://tikz.org/chapter-15`. The GitHub link for downloading is `https://github.com/PacktPublishing/LaTeX-graphics-with-TikZ/tree/main/15-fun`.

This chapter will utilize the following packages: `tikzducks`, `tikzlings`, `bearwear`, `scsnowman`, `tikzpingus`, `tikzpeople`, `jigsaw`, `tikzbricks`, `tikz-3dplot`, and `worldflags`.

Remember, you must always load a package with `\usepackage` to be able to use it.

The package names speak for themselves, so buckle up; it will be a rollercoaster ride!

Drawing cute creatures

The internet is full of memes about animals and cartoon characters. They have found their way into TikZ as well. In addition, avatars of users from the TikZ community and their favorite animals have been immortalized through the TikZ code. Let's see some examples.

Playing with rubber ducks

Ducks are well-known for internet memes. In the developer community, for example, **rubber duck debugging** is a famous software debugging technique where the programmer explains their code in front of a rubber duck in every detail. By explaining it and articulating details and reasons, programmers can identify issues or bugs they would not have noticed by staring at the code. The rubber duck toy is also here rather as a meme than the solution itself. You can use, for example, a teddy bear instead, whom you will meet in the next section.

Rubber ducks, as classic toys, are so playful and fun that they have become a popular collectible item. Companies offer rubber ducks with their brand design to customers as promotional merchandise. For example, I work for Lufthansa, and our merchandise includes rubber ducks dressed as pilots or flight attendants. Overleaf has rubber ducks with their brand design.

Long story short, there's the `tikzducks` package for drawing rubber ducks with many variations in design so that a duck can represent a famous person, a profession, a county, a hobby, or a cliché. It was invented and maintained by samcarter, who added many features over time. You met the ducks already in *Figure 9.1* and *Figure 11.4*.

The ducks can be customized with color and text for hair, clothing, and accessories. The package manual contains a huge list of options. Just browse the manual, make your choice, and add an option. Here's how I went through and made some quick choices:

```
\duck[crazyhair = brown!60!black, glasses, eyebrow,
  signpost = TikZ, speech = Use it!, laughing,
  jacket = orange, lapel, buttons, water]
```

When you use this command in a `tikzpicture` environment, you get the following design:

Figure 15.1 – A duck in water with flashy clothes and a signpost

To give an impression of the variety of choices, here's a bunch of duck designs where I picked various options. I used a `matrix` node to easily position them in the following code:

```
\node [matrix] {
  \duck[laughing, tophat, bowtie=violet, jacket=black,
    buttons=violet, recedinghair=black!80,
    wine, eye=red!40] &
  \duck[magichat, recedinghair=lightgray,
      jacket=violet, beard=lightgray, magicwand] &
  \duck[parrot, stethoscope=black!70, jacket=gray!30,
      buttons=gray, squareglasses, longhair=gray] \\
  \duck[snowduck=lightgray!60] &
  \duck[umbrella=red!70, handbag=red, bill=red!70,
    jacket=pink!80!black, longhair=yellow,
    necklace=magenta, sunglasses=magenta] &
  \duck[alien, laughing, body=green!70!black,
    bill=green!50!black, crystalball] \\ };
```

This gives us six individual ducks:

Figure 15.2 – Ducks with various styles and accessories

You can start playing with colors or choose among the many options and accessories in the manual that you can find at `https://texdoc.org/pkg/tikzducks`.

You can find many more examples on GitHub at `https://github.com/samcarter/tikzducks`, such as dressed as famous people such as Donald E. Knuth, the Queen, Shakespeare, Super Mario, Mr. Spock, and representing various professions with clothing and accessories.

That package was the first of its kind. More creatures appeared; let's meet them in the next section.

Meeting the TikZlings

samcarter wrote the `tikzlings` package, a collection of cute animals and other characters drawn in TikZ. Fun trivia, I suggested this name on August 27, 2018, in the TeX StackExchange chat, and **TikZlings** was chosen as the name for the creatures and the package.

The package stays a work in progress, as new TikZlings, features, and accessories may be added anytime. In 2023, we can find 24 inhabitants. Most easily, it's displayed by a single command: you write `\chicken`, and you get a chicken.

Here are birds and insects:

Figure 15.3 – TikZlings: \owl, \chicken, \penguin, \bee, and \bug

Then we got some furry friends:

Figure 15.4 – TikZlings: \squirrel, \marmot, \moles, \sloth, \pig, \koala, \coati, \panda, \cat, \mouse, \sheep, and \wolf

And this is the rest of the team:

Figure 15.5 – TikZlings: \elephant, \hippo, \rhino, \anteater, \bat, \snowman, and \bear

You can use any standard TikZ options, such as here:

```
\penguin[owl=20, xshift=2cm, scale=0.5]
```

All TikZlings can be displayed from behind, too, in a three-dimensional appearance, and outline only so that a kid can color it. For example, this is what the pig can look like:

Figure 15.6 – Variations of drawing a TikZling: \pig, \pig[back], \pig[3D], and \pig[contour]

You can set further options such as `body=green` and `eye=blue`, and you can add accessories, just like with `tikzducks`, such as hats, books, and signposts. See *Figure 15.24* for an example of this.

At `https://texdoc.org/pkg/tikzlings`, you can find the manual with all available options.

At `https://github.com/samcarter/tikzlings`, you can find the source code and examples of modifications. At `https://github.com/TikZlings`, you can find videos and more source code.

Apart from about 50 accessories such as handbag, umbrella, broom, shovel, and food such as cake, pizza, baguette, cheese, and various hats, there's additional styling for the teddy bear: the `bearwear` package by Ulrike Fischer provides some shirts with design options. After drawing a teddy bear, call the `\bearwear` command with your desired options. In the following example, I chose a shirt in yellow and red, with a TikZ sign on it, and a picture of his friend, the bat:

```
\bear
\bearwear[shirt={shade, top color=yellow,
    bottom color=red}, body deco={\node[scale=0.5]
    at ([yshift=0.8mm]bearheart) {\tiny TikZ};
    \pic at (beartummy)[scale=0.18, yshift=-1cm]{bat};} ]
```

Now the bear is wearing a T-shirt:

Figure 15.7 – A teddy bear wearing a T-shirt

Again, you have a lot of customization options that you can read in the manual at https://texdoc.org/pkg/bearwear.

You saw a snowman in *Figure 15.5*. Did you know that snowmen in Japan are different? You will see that in the next section.

Building snowmen

In Japan, traditionally, snowmen are made from two big snowballs. They don't have a middle section like the western snowmen and usually have no arms. Fortunately, a package to draw them in TikZ was written by Hironobu Yamashita, a maintainer of Japanese LaTeX variants.

If you load the `scsnowman` package, the `\scsnowman` command draws a basic snowman outline that you can see in *Figure 15.8* on the left side. The command uses TikZ internally, so you don't need a `tikzpicture` environment.

Initially, the snowman is pretty small, so you may want to scale it, by using `\scsnowman[scale=2]`. Many more options and colors are supported, as with the ducks and the TikZlings packages. Here are two quick examples:

```
\scsnowman[hat, arms, buttons, snow=blue, note=red]
\scsnowman[arms, muffler=red, hat=blue, broom=brown]
```

And here's how they look compared to the basic shape:

Figure 15.8 – Snowman variations

You can find the entire feature documentation at `https://texdoc.org/pkg/scsnowman`.

Apart from ducks, teddy bears, and snowmen, penguins are popular, too, so there's a dedicated package.

Playing with penguins

The `tikzpingus` package by Florian Sihler got its name from the German spelling "pinguin" for that Antarctic bird, or **pingu** for short. It was inspired by `tikzducks`, especially regarding the wealth of features. Let's have a quick look. Use the `\pingu` command in a `tikzpicture` environment to draw the basic penguin. It will look like this:

Figure 15.9 – A penguin

Choose a few options from the manual and change the appearance or add extras, such as here:

```
\pingu[eyes shiny, crown, gold medal, right wing wave]
```

The penguin now looks as follows:

Figure 15.10 – A penguin with accessories

The manual at https://texdoc.org/pkg/tikzpingus is over 120 pages long and covers everything regarding coloring, wing positions, and accessories.

Let's turn from animals to humans now.

Picturing people

There's a package designated to drawing human creatures. Nils Fleischhacker designed the `tikzpeople` package initially to depict cryptographic protocols between parties. It should visualize humans and their communication flows.

He quickly added many shapes representing various kinds of people and professions. You saw an example in *Figure 3.9*.

These are node shapes, so you can add any node and give it the desired shape, as shown here:

```
\node[businessman, minimum size=2cm] at (2,1) {};
```

You can add text to the node as usual if you like. The `minimum size` key defines how big the node will become. The node has many predefined anchors; you can see example anchors here, both with names and angles:

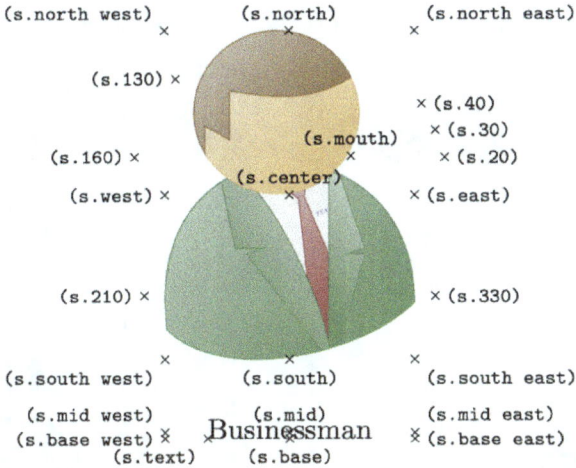

Figure 15.11 – A human shape

You can use the anchors to connect speech bubbles or any accessory, arrows, or curves. First, we draw the shape of Agent K. from the *Men in Black* movie:

```
\node[name=k, shape=maninblack, minimum size=1cm] {};
```

Then we add his signature phrase, using a callout shape positioned near the mouth node:

```
\node[ellipse callout, xshift=1.2cm, draw,yshift= .3cm,
  callout absolute pointer={(k.mouth)},
  font=\tiny\sffamily, align=center, inner sep=1pt]
  {Please,\\stand back.};
```

The output is the following:

Figure 15.12 – A human shape example

29 base shapes are currently available. The first twenty shapes are shown in *Figure 15.13*:

Figure 15.13 – A variety of human shapes

The remaining 9 shapes are as follows:

Figure 15.14 – More human shapes

You can modify colors and add accessories, as we did for *Figure 3.9*. The list of all available options is in the manual, which you can find at https://texdoc.org/pkg/tikzpeople.

Now, let's play some games.

Playing and crafting

We can use TikZ for displaying or documenting games. Here are some example packages:

- The `logicpuzzle` package can print puzzles such as Sudoku and Minesweeper
- The `JeuxCartes` package can display playing cards, such as for Poker, Tarot, and Uno
- The `rubikcube` package provides commands and macros for typesetting Rubik's cube configurations, rotation sequences, and move notation
- The `havannah` package prints diagrams of the board games Havannah and Hex

The internals of TikZ are often hidden within the package source code. Sometimes they are not really visible to the user because the packages provide their own syntax. Since we are learning TikZ, we pick two games where we use commands within TikZ; we will choose jigsaw puzzles and building with **bricks**, as you know from **Lego**.

Creating jigsaw puzzles

We all know **jigsaw** puzzles: a set of flat shapes are assembled until they form a picture such as a photo or drawing. The `jigsaw` package can draw adjustable puzzle pieces, combine them, and even create complete and randomized puzzles for you.

You can draw a single piece with the following command:

```
\piece[red]{-1}{-1}{0}{0}
```

It has four arguments for the bottom, right, top, and left side, in the following order: `1` means a slot, `0` stands for a straight line, and `-1` creates a tab. This red piece is displayed in *Figure 15.15*.

You can use scopes to move pieces like this:

```
\begin{scope}[shift={(1,0)}]
  \piece[yellow]{-1}{0}{0}{1}
\end{scope}
```

Here, the yellow piece is moved by (1,0), which means by 1 in the *x* direction and by 0 in the *y* direction. Such shifting is also displayed in *Figure 15.15* when you look at the yellow piece.

The default width and height of the pieces is 1, and you can use shifting or positioning to build a puzzle.

A `matrix` may be easier to type once we find a good value to subtract as row and column separation to move the pieces together. We can do it like this:

```
\matrix[column sep=-0.315cm, row sep=-0.315cm] {
  \piece[red, overlay]{-1}{-1}{0}{0}
& \piece[yellow]{-1}{0}{0}{1} \\
  \piece[blue]{0}{-1}{1}{0}
& \piece[green]{0}{0}{1}{1} \\ };
```

This gives us the following output:

Figure 15.15 – A mini jigsaw puzzle

There's a command to create larger puzzles. The \jigsaw{10}{6} command creates a 10x6 puzzle as follows:

Figure 15.16 – A bigger jigsaw puzzle

You can create a puzzle game with a jigsaw puzzle pattern over photos or drawings. In the following code, we fill an area with gray color, add a picture, and then draw the jigsaw pattern over it:

```
\fill[gray!40] (0,0) rectangle (3,3);
\pic[scale=1.25] at (1.5,0) {chicken};
\jigsaw{3}{3}
```

This gives us a small 3x3 puzzle with an image:

Figure 15.17 – A jigsaw puzzle with a picture

Now, we could print it on paper, glue the paper to cardboard for stability, and cut the puzzle pieces along the curves with a sharp craft knife on a cutting mat. Then, shuffle and have fun!

Building with bricks

Many kids love building with bricks, and so do I. The `tikzbricks` package, also written by samcarter, is for displaying bricks and building shapes from them. It looks like Lego, but it is not called by that commercial name.

Once you have loaded the package, you can draw a single brick in a `tikzpicture` environment using a `\brick` command with two arguments: the block's length and width. An optional argument is for the color.

See, for example, *Figure 15.18* where we use `\brick[color=red]{1}{3}` and `\brick[color=blue]{3}{2}`:

Figure 15.18 – Colored bricks

We can place the bricks as we like; we should just be aware that we draw them in order from back to front, so we see the pieces in front, while bricks in the back can be partially hidden and not the other way around.

To make building walls easier, a `wall` environment is provided for placing bricks next to each other in an easier way:

- Everything should be drawn from right to left, and from bottom to top, for proper viewing. No surrounding `tikzpicture` environment is needed.
- Use the `\wallbrick` command instead of `\brick`, with the same options.
- Adjust the `brickx` or `bricky` counters to have a gap in the *x* or *y* direction.
- Use the `\newrow` command to go up to the next row and back to the start at the right-hand side.

The following code builds a wall:

```
\begin{wall}
  \wallbrick[color=blue]{2}{2}
  \wallbrick[color=red]{1}{2}
  \stepcounter{brickx}
  \wallbrick[color=green]{2}{2}
  \newrow
  \stepcounter{brickx}
  \wallbrick[color=brown]{4}{2}
  \newrow
  \addtocounter{brickx}{2}
  \wallbrick[color=orange]{2}{2}
\end{wall}
```

The output is the following:

Figure 15.19 – A wall of colored bricks

I have to admit that I was tempted to play with it. I remembered the `pxpic` package by Jonathan P. Spratte, which creates pixel-art images. The manual contains a pixel image of Tux, the Linux mascot:

Figure 15.20 – A pixel image of Tux, the penguin

Based on this, I manually built the following bricks wall, row by row:

Figure 15.21 – Tux, the penguin, built with bricks

The source code is available at https://tikz.org/tux.

Internally, the drawing is a 3D plot done with the `tikz-3dplot` package. You can use commands from that package to modify the viewing angle. For example, with `\tdplotsetmaincoords{120}{-60}` the output changes:

Figure 15.22 – A different viewing angle

`\tdplotsetmaincoords{70}{160}` is the default. For more information, take a look at the `tikz-3dplot` manual at https://texdoc.org/pkg/tikz-3dplot.

The source code for this image is also on `TikZ.net` and GitHub.

The building process can be automated. Scott Pakin wrote a Python script that converts PNG images to `tikzbricks` code. You can download it at https://github.com/samcarter/TikZbricks/blob/main/img2bricks. It is also described in the package manual. You can find the `tikzbricks` manual at https://texdoc.org/pkg/tikzbricks.

After the games, let's finally look at an application that can be used for international competitions such as the Olympic games or sports world cups: drawing flags of nations.

Drawing world flags

For use in LaTeX documents and particularly in TikZ drawings, Wilhelm Haager created the `worldflags` package, which contains the national flags of all independent nations. Where possible, he used TikZ to draw them with geometric constructions; otherwise, he converted **Scalable Vector Graphics** (**SVG**) files via Inkscape to TikZ.

The main command is as follows:

```
\worldflag[options]{country code}
```

Here, `country code` is the common two-letter code for the country, such as `US` for the United States or `NZ` for New Zealand. The manual provides the complete list.

These are some of the supported options:

- `width` and `length` stand for the dimensions of the flag. With `length` of zero, the default aspect ratio is used for calculating the final length
- `framewidth` is the line width of the frame
- `framecolor` sets the frame color
- `emblem` or `noemblem` enables or disables the drawing of an emblem on the flag

So, for example, we can write the following code:

```
\worldflag[width=2cm, framewidth=0.3mm,
  framecolor=black]{BR}
```

This gives us the flag of Brazil with a width of 2cm – it may be scaled a bit here in print or display:

Figure 15.23 – The Brazilian flag

When we use several flags, such as for displaying international sports results, we can use \flags default[options] to set values for all flags at once and then simply use \worldflags {country code} in the document.

As TikZ users, we can use a pic element in our drawings. This can be done as follows:

```
\pic (id) [country=XX, ...] at (x,y) {worldflag};
```

Here, id is the name of the pic element for reference; XX is again the country's two-letter code, and (x,y) is a coordinate. We can use the same options as here, plus the following:

- hang is a value in degrees that indicates hanging down from a flagpole
- turn can be in degrees, and rotates the flag around the imaginary flagpole
- rotate, scale, xshift, yshift, and other standard TikZ transformation options can also be used

We are in the fun chapter, so let's have a fancy example. First, we draw a hippo from the tikzlings package with a thought bubble and a toy lightsaber serving as a flagpole:

```
\hippo[body=brown!60!black, lightsaber=brown,
  think={\textsf{The end}}]
```

Our hippo lives in Malawi, so we draw the flag of Malawi, put it next to the flagpole, scale it by 50%, rotate it by 30 degrees, and let it hang down by 20 degrees:

```
\pic [country=MW, scale=0.5, rotate=-30, hang=20]
  at (1.02,1.1) {worldflag};
```

Our book now ends with a hippo with a lightsaber and his home flag:

Figure 15.24 – A hippo with the flag of Malawi

I hope that you enjoyed such a playful chapter after so much learning. It's exciting to see what the TikZ community created and what you can do yourself now.

Index

Symbols

3D plotting 220-222

A

anchors 28
angles
  adding 168-170
arc 20
area
  filling 102
  shading 117
area between plots
  filling 213-215
arguments 71
  adding, to styles 71-73
arrows
  drawing 56, 57
arrow tips
  barbed arrow tips 57
  customizing 58, 59
  geometric arrow tips 58
  mathematical arrow tips 57
Asymptote 3, 204
  reference link 3
axis environments 207-209
axis shading 117-120

B

background layer
  drawing 153, 154
ball shading 121
barbed arrow tips 57
bar chart 243-247
Bézier curves 195
Bézier splines
  using, to connect point 196
bilinear interpolation shading 121, 122
bricks building game
  creating 266-269
bubble diagram 232
bullet list 237

C

canvas
  transforming 187, 188
Cartesian coordinates 13-15
CentOS 6
circle 20

circular flow diagrams 231
circular relationship diagrams 232
clipping 110
clipping area 110
clipping path 110
color maps 220
colors
  using 21, 22
color wheel shading 122-124
comma-separated values (CSV) 240
connected constellation diagram 233
constellation diagrams 233
coordinate axes
  creating 206, 207
  fine-tuning 206, 207
coordinates
  adding 165, 166
  angles, adding 168-170
  calculating with 165
  Cartesian coordinates 13-15
  points, computing between 166, 167
  polar coordinates 15, 16
  projection on line 167, 168
  relative coordinates 18, 19
  shifting 178, 179
  subtracting 165, 166
  three-dimensional coordinates 16-18
  units 19, 20
  working with 12
creatures
  drawing 254
  penguins, playing with 259
  people, picturing 260-263
  rubber ducks, playing with 254, 255
  snowmen, building 258
  TikZlings 256-258

cubic Bézier curves
  specifying 195, 196
curve
  creating, manually through point 190-193
cycle list 212

D

Debian 5
decorations 129-132
  adjusting 139-142
decoration types
  exploring 132
  markings, adding 138, 139
  morphing paths 132-134
  paths, decorating with text 137
  paths, replacing with ticks 135, 136
descriptive diagram 237
  writing 237-239
distance modifier 167
domain 209
drawing
  clipping 110-113

E

edge options 54
  connection options 55, 56
  path options 55
edges 49
  text, adding to 51-54
ellipse 20
Encapsulated PostScript (EPS) 45
epstopdf 45
even odd rule 107
  pros 108
  versus nonzero rule 108, 109

F

Fedora 6
flowchart
  creating 226
flow diagrams 226
  circular flow diagrams 231
  linear flow diagrams 226-230
Font Awesome icon library 237
foreground layer
  drawing 153, 154
for loop 162

G

games
  bricks, building 266-269
  documenting 263
  jigsaw puzzles, creating 264-266
geometric arrow tips 58
geometric shapes
  drawing 20, 21
GNU Octave 204
GNUplot 204
graphs
  producing 91-94
grid 20

H

handlers 64
havannah package 263
Hobby algorithm
  using, to connect point 197-201
human shape
  picturing 260-263

I

icons 44
images
  inserting, into nodes 44-46
intersection library
  keys and options 173
intersections of paths
  calculating 172-175
isometric projection 16

J

JeuxCartes package 263
jigsaw puzzles
  creating 264-266

K

key handler 73
keys
  example 63
Koch curve 141
Koch snowflake 140

L

labels
  customizing 209-211
LaTeX content
  overlaying, with TikZ drawings 155, 156
LaTeX picture environment 2, 3
layers of image 145
legend 216
  adding 216
Lego 264
line 20
  nodes, placing along 38, 39

Index

linear flow diagrams 226-230
line chart 240-243
logicpuzzle package 263
loop repetitions
  counting 170
loops
  repeating 162-165
loop variables
  evaluating 170, 171
  remembering 171, 172

M

MacTeX 5
Mathematica 204
mathematical arrow tips 57
Matlab 204
Matplotlib 204
matrix
  positioning in 95-98
MetaPost 3, 197, 204
  reference link 3
MiKTeX 5
mind maps 88
  creating 88-91
minor ticks 210
module 227, 229
morphing paths 132-134

N

named paths 173
Network Topology Icons
  reference link 45
node chain 92

nodes 25, 26, 49
  aligning 36
  aligning, at text baseline 39, 40
  anchor 28
  anchors positioning, used 36-38
  anchors, using 28
  connecting, by edges 50, 51
  images, inserting into 44-46
  labels, adding 43, 44
  pins, adding 43, 44
  placing, along line 38, 39
  positioning 36
  relative positioning, used 36-38
  shapes 26, 27
  shapes, using 28
  shifting 178, 179
  spacing, within and around 34-36
  with arrows 28
  with colors 26
node text baseline
  pictures, aligning 40-42
nonzero rule 103-106
  pros 108
  versus even odd rule 108, 109

O

operating system TeX installation
  used, for installing TikZ 5
origin 13

P

parametric plotting 218-220
partway modifier 166
path 12, 55
  using, multiple times with pre- and post-actions 128, 129

path interior 102
  even odd rule 107
  nonzero rule 103-106
path morphing 131
penguins
  playing with 259
pics
  creating 73-76
  using 73-76
pictures
  aligning, at node text baseline 40-42
pie chart 247-249
pingu 259
pitch 181
planet node 233
plot
  producing, advantages in
    LaTeX with TikZ 204
  using, to connect point 193-195
plot intersections
  calculating 215, 216
plotting 204, 205
  commands, using 211, 212
  options, using 211, 212
polar coordinates 15, 16
  using 217, 218
Portable Graphics Format (PGF) 2
  reference link 6
pre- and post-actions
  for using, path multiple times 128, 129
priority descriptive diagram 238
projection modifier 167
projection on line 167, 168
PStricks 4, 204
  URL 4
Python 204

Q

quantitative diagrams
  bar chart 243-247
  line chart 240-243
  pie chart 247-249
  producing 239
  wheel chart 249, 250
quotes syntax 51

R

R 204
radial shading 120
radians 218
rectangle 20
rectangular grid 10, 11
Redhat 6
relationship diagrams
  building 232-236
relative coordinates
  using 18, 19
remember picture option
  exploring 157-159
reverse clipping 114-117
roll 181
rotating 179-181
  options 179, 181
rotation 179
rubber duck debugging 254
rubber ducks 254
  creating 254, 255
rubikcube package 263

S

satellite nodes 233
Scalable Vector Graphics (SVG) 109, 270

scaling 181-184
 options 181
segment 215
sequence diagram 229
Seven Bridges of Königsberg 65
shading 22
shapes, nodes 31-34
 circle shape 30
 coordinate shape 30
 ellipse shape 30
 example collection 32
 rectangle shape 29
 reference link 34
shifting 178
 options 178
shipout 158
slanting 185-187
 options 185
snowflake 142
snowmen
 building 258
 reference link 259
splines 196
stacked bar chart 244
stacked line chart 242
stencils 44
styles 63
 arguments, adding 71-73
 defining 64
 graph example 65
 inheriting 68, 69
 using 64-68
 using globally 69, 70
 using locally 69, 70

T

TeX Directory Structure (TDS) 6
TeX Live 5
text
 adding, to edges 51-54
text baseline
 nodes, aligning 39, 40
third-party software
 using, drawbacks 204
three-dimensional coordinates 16-18
ticks
 customizing 209-211
tikz-3dplot manual
 reference link 270
tikzbricks manual
 reference link 270
TikZ documentation
 reference link 7
 working with 6, 7
TikZ drawings
 used, for overlaying LaTeX content 155, 156
TikZ figure
 creating 7, 8
TikZ ist kein Zeichenprogramm (TikZ) 2
 benefits 4, 5
 installing 5
 installing, from sources 6
 installing, with vanilla TeX distribution 5
 key 64
 reference link 172
 with operating system TeX installation 5
TikZlings 75, 256-258
 reference link 257
tikzpeople package
 reference link 33

tikzpicture environment
   using 10-12
to operation
   using 59-61
transparency
   using 146-153
transparency group 150
transpose of matrix 151
trees
   drawing 80-88

U

Ubuntu 6
units
   using 19, 20

V

vanilla TeX distribution
   used, for installing TikZ 5
variable 162

Venn diagram
   drawing, to display intersections of set 147
vertices 65

W

wheel chart 249, 250
   reference link 250
world flags
   drawing 270-272
worldflags package 270

Y

yaw 181

www.packtpub.com

Subscribe to our online digital library for full access to over 7,000 books and videos, as well as industry leading tools to help you plan your personal development and advance your career. For more information, please visit our website.

Why subscribe?

- Spend less time learning and more time coding with practical eBooks and Videos from over 4,000 industry professionals
- Improve your learning with Skill Plans built especially for you
- Get a free eBook or video every month
- Fully searchable for easy access to vital information
- Copy and paste, print, and bookmark content

Did you know that Packt offers eBook versions of every book published, with PDF and ePub files available? You can upgrade to the eBook version at www.packtpub.com and as a print book customer, you are entitled to a discount on the eBook copy. Get in touch with us at customercare@packtpub.com for more details.

At www.packtpub.com, you can also read a collection of free technical articles, sign up for a range of free newsletters, and receive exclusive discounts and offers on Packt books and eBooks.

Other Books You May Enjoy

If you enjoyed this book, you may be interested in these other books by Packt:

LaTeX Cookbook

Stefan Kottwitz

ISBN: 9781784395148

- Choose the right document class for your project to customize its features
- Utilize fonts globally and locally
- Frame, shape, arrange, and annotate images
- Add a bibliography, a glossary, and an index
- Create colorful graphics including diagrams, flow charts, bar charts, trees, plots in 2d and 3d, time lines, and mindmaps
- Solve typical tasks for various sciences including math, physics, chemistry, electrotechnics, and computer science
- Optimize PDF output and enrich it with meta data, annotations, popups, animations, and fillin fields
- Explore the outstanding capabilities of the newest engines and formats such as XeLaTeX, LuaLaTeX, and LaTeX3

LaTeX Beginner's Guide - Second Edition

Stefan Kottwitz

ISBN: 9781801078658

- Make the most of LaTeX's powerful features to produce professionally designed texts
- Download, install, and set up LaTeX and use additional styles, templates, and tools
- Typeset math formulas and scientific expressions to the highest standards
- Understand how to include graphics and work with figures and tables
- Discover professional fonts and modern PDF features
- Work with book elements such as bibliographies, glossaries, and indexes
- Typeset documents containing tables, figures, and formulas

Packt is searching for authors like you

If you're interested in becoming an author for Packt, please visit `authors.packtpub.com` and apply today. We have worked with thousands of developers and tech professionals, just like you, to help them share their insight with the global tech community. You can make a general application, apply for a specific hot topic that we are recruiting an author for, or submit your own idea.

Share your thoughts

Now you've finished *LaTeX Graphics with TikZ*, we'd love to hear your thoughts! Scan the QR code below to go straight to the Amazon review page for this book and share your feedback or leave a review on the site that you purchased it from.

`https://packt.link/r/1804618233`

Your review is important to us and the tech community and will help us make sure we're delivering excellent quality content.

Download a free PDF copy of this book

Thanks for purchasing this book!

Do you like to read on the go but are unable to carry your print books everywhere?

Is your eBook purchase not compatible with the device of your choice?

Don't worry, now with every Packt book you get a DRM-free PDF version of that book at no cost.

Read anywhere, any place, on any device. Search, copy, and paste code from your favorite technical books directly into your application.

The perks don't stop there, you can get exclusive access to discounts, newsletters, and great free content in your inbox daily

Follow these simple steps to get the benefits:

1. Scan the QR code or visit the link below

`https://packt.link/free-ebook/9781804618233`

2. Submit your proof of purchase
3. That's it! We'll send your free PDF and other benefits to your email directly

www.ingramcontent.com/pod-product-compliance
Lightning Source LLC
Chambersburg PA
CBHW080730300426
44114CB00019B/2533